Earth, Wind and Fire:

Unpacking the Political, Economic and Security Implications of Discourse on the Green Economy

This research project was supported by:

Earth, Wind and Fire:

Unpacking the Political, Economic and Security Implications of Discourse on the Green Economy

Eds. Lynn Krieger Mytelka, Velaphi Msimang and Radhika Perrot

Mapungubwe Institute for Strategic Reflection (MISTRA)
First floor, Cypress Place North
Woodmead Business Park
142 Western Service Road
Woodmead 2191
Johannesburg

First published October 2015

ISBN 978-1-920655-09-9

Published by Real African Publishers
on behalf of the Mapungubwe Institute for Strategic Reflection (MISTRA)

PO Box 3317
Houghton
Johannesburg 2041

Sub-Editor: Barry Gilder
Copy editor: Angela McClelland
Indexer: Jackie Kalley

MAPUNGUBWE INSTITUTE (MISTRA)
[A NON-PROFIT COMPANY][104-474-NPO]
REGISTRATION NUMBER 2010/002262/08
["THE INSTITUTE"]

Preface

While there may be continuing debate about magnitude, impact and implications, there is universal consensus on the Anthropocene: an epoch in which we currently live, where human action is having a significant impact on the earth's environment.

Critical questions about a united global effort now earnestly engage the minds of decision-makers across the public and private spheres, and in civil society. A systematic partnership across these sectors is emerging, with scientists playing a crucial role in outlining the challenges and defining the interventions required to prevent an existential disaster for humanity.

Yet, reaching agreement on concrete action for each of the role-players on a worldwide scale has been a global imbroglio of stops and starts. What are the reasons behind this? Perhaps contestation around the science may have something to do with it. And so may the finger pointing and arguments on common but differentiated responsibility; or the rancour on funding sources for adaptation and mitigation. The reasons are many and varied.

Through this book, the Mapungubwe Institute for Strategic Reflection (MISTRA) seeks to dig below the surface of these debates. While examining some of the obvious reasons behind the spasmodic progress in addressing global warming, it focuses on the deeper self-interests that inform the postures of stakeholders. The treatise pays particular attention to South Africa's own tentative steps, along with its neighbours, in contributing to these efforts.

The title, *Earth, Wind and Fire: Unpacking the Political, Economic and Security Implications of Discourse on the Green Economy*, borrows from the name of an American band (EWF) that has, over four-and-half decades, dazzled the world with its outstanding talents in a variety of musical genres.

But it is from more than the band name that the researchers have drawn inspiration. As with EWF's versatility, the central theme of this work is that discourse and policy action on the Green Economy should fully embrace a transdisciplinary approach. For any policy to be meaningful and effective, it should acknowledge social complexity and take on board the multidimensional nature of the issues. This rings even truer for green policies, which require even higher levels of policy coordination and integrated governance. At the same time, the desire to move with speed needs to be tempered by a comprehensive understanding of the path dependencies that bear down on specific economies and polities.

Beyond these macro-issues are matters of day-to-day praxis: the minute care required in building community partnerships for specific local interventions; the curve balls of South Africa's current double jeopardy in the energy-water nexus, and the opportunities that programmes such as low-income housing development can present.

The book does acknowledge the impressive progress that South Africa has made in its procurement process for renewable energy. Yet, it also warns that long-term sustainability of such programmes depends also on maximising socio-economic impact, especially among communities previously marginalised in the country's historical socio-economic evolution.

The development of *Earth, Wind and Fire* was not confined merely to the writing of the chapters of this book. It was also extended to engagements with stakeholders in public discourses on current issues such as the stark energy choices South Africa needs to make in the light of current power shortages. In addition, workshops were held both in South Africa and France around preparations for the Paris COP21 event later in 2015.

There are many more themes that research work of this nature could have covered. But it is the hope of the authors that the various chapters help to illustrate the complex issues in transitioning to a green economy. The researchers would have more than met their ambition if this work contributes to ensuring a better understanding of the challenges associated with that transition. Better still, if the policy recommendations contained in this book do help to inform the framing of future courses of action among the stakeholders in South Africa and further afield, the team would have been honoured to add a small brick in the wall of sustainable development.

This research project was undertaken with the financial support of the Development Bank of Southern Africa (DBSA), which manages the Green Fund on behalf of the Department of Environmental Affairs, and of the Friedrich-Ebert-Stiftung (FES) South Africa Office. For this, and for the hard work of the authors, the peer reviewers, the editors and the support staff, MISTRA is profoundly grateful.

Joel Netshitenzhe
Executive Director

Acknowledgements

The Mapungubwe Institute for Strategic Reflection (MISTRA) would like to express its deepest gratitude to the project leader of this research project, Professor Lynn Mytelka; the MISTRA faculty head who supervised the project, Dr Velaphi Msimang; the project co-coordinator, Radhika Perrot; and the researchers for consenting to join the project and undertaking the work.

Appreciation is also extended to subject specialists who provided invaluable comments, insights and feedback, thus strengthening the book's content and analyses: Professor Anthony Black (University of Cape Town), Christopher Crozier (REDISA), Dr Ziboneni Godongwana (REDISA), Dhesigen Naidoo (Water Research Council), Dr Lucy Baker (Sussex University, United Kingdom), Jens Pedersen (Médecins Sans Frontières), Professor Nathalie Lazaric (CNRS, France), Richard Doyle (Linkd Environmental Services), Richard Worthington (Project 90x2030), Holle Wlokas (University of Cape Town) and Wikus Kruger (University of Cape Town).

A number of MISTRA staff contributed to the successful outcome of this project. They include: Xolelwa Kashe-Katiya for liaising with the funding organisation on project and financial updates and reporting; Lorraine Pillay, who expedited contracts and invoices; Wilson Manganyi, Duduetsang Mokoele, Khangwelo Tshivhase, Hope Prince, Asanda Luwaca, Thabang Moerane, Pearl Sesoko and Rose Ngwato for their substantive and logistical support. Appreciation is also extended to Joel Netshitenzhe for meticulously reading and editing the manuscript; to Siphokazi Mdidimba for ensuring the chapters complied with the MISTRA Style Guide and to Barry Gilder for sub-editing the manuscript.

MISTRA also extends its thanks to Angela McClelland and Reedwaan Vally of Real African Publishers who copy-edited, designed and produced this publication and to Robert Machiri for the design of the book cover.

MISTRA warmly thanks its partner organisation, the Development Bank of Southern Africa (DBSA), in particular Michelle Layte and Nomsa Zondi for their assistance and supportive interaction over the duration of this project

MISTRA warmly thanks its donors who contributed to this project as well as those who have generally contributed to the work of the Institute.

Project Funders

- Development Bank of Southern Africa
- Friedrich-Ebert-Stiftung

MISTRA Funders and Donors

SA Corporates

- Absa
- Anglo American
- AngloGold Ashanti
- Anglo Platinum
- Aspen Pharmacare
- Batho Batho Trust
- Brimstone
- Chancellor House
- Discovery
- Encha Group Limited
- Kumba Iron Ore
- MTN
- Mvelaphanda Management Services
- Naspers
- Power Lumens Africa
- Safika
- Sanral
- Sasol
- Shanduka Group
- Simeka Group
- Standard Bank
- Yard Capital Development Trust
- Yellowwoods

Local Foundations

- First Rand Foundation
- Nedbank Foundation
- Oppenheimer Memorial Trust
- Social Science Development Forum (SOSDEF)
- Transnet Foundation

International Sources

- Embassy of the People's Republic of China

Individuals

- Robinson Ramaite
- Thandi Ndlovu

Partners

- Lilliesleaf
- National Advisory Council for Innovation (NACI)
- Nelson Mandela Foundation
- South African Reserve Bank
- University of Johannesburg
- University of Pretoria
- University of South Africa
- University of Witwatersrand

Contents

List of Contributors

Radhika Perrot is a Senior Researcher in the Knowledge Economy and Scientific Advancement (KESA) Faculty at the Mapungubwe Institute for Strategic Reflection (MISTRA) in Johannesburg, South Africa. Her research interest is in science, technology and innovation policies (S&T&I) of low-carbon energy technologies, covering issues and topics around socio-technological factors of renewable energy technologies and innovation and market competition and firm strategies in local and global markets. Prior to moving to South Africa, she worked in India, the Netherlands and Belgium researching and writing on topics around renewable energy development. At MISTRA, she is researching and project coordinating a number of projects such as the political and economic discourse on the green economy in South Africa; the socio-economics and ethics of nano- and biotechnologies; the curse of epidemics and the socio-psychological, economic and political implications; low-carbon innovation in Africa, and the politics of nuclear in South Africa. She is a Ph.D. Candidate at the United Nations University (UNU-MERIT) – University of Maastricht, the Netherlands, and a Research Fellow at the Institute of Economic Research on Innovation (IERI) of the Tshwane University of Technology in Pretoria.

Manisha Gulati is an Energy Economist with the World Wide Fund for Nature (WWF) South Africa. She has over 14 years of multidisciplinary experience drawing on development, resource and environmental economics covering India and South Africa. Her areas of expertise are energy, low-carbon development, the green economy and resources nexus. She specialises in quantitative and qualitative analysis, and policy formulation and advisory. At WWF South Africa, Manisha leads the work on energy, low-carbon transition and green economy for South Africa. She co-manages the food-energy-water nexus programme. Manisha started her career at The Energy and Resources Institute (TERI), India's leading energy and environment focused research organisation and has since worked in the consulting and financing sectors. She is an economics graduate from St Stephen's College, India, and holds a Master's Degree in Business Economics from the University of Delhi, India. She has several publications to her credit.

Louise Scholtz has an MBA, an M.Phil. in Sustainable Development and Planning, and is Manager in the Living Planet Unit at WWF-SA. Her focus areas include research relating to energy policy and infrastructure, sustainable strategies and development, and the green economy. She is also the lead on the following WWF-SA programmes/projects: WWF China/Africa Programme – with a particular focus on engagement with the Forum on China/Africa Cooperation (FOCAC); Climate Solver; 'Greening of Social Housing' project – partnering with the National Association of Social Housing Organisations (NASHO) and Communicare; and the Sustainable Public Procurement Project – partnering with the International Institute for Sustainable Development (IISD) and the Government of the Western Cape. Given her prior legal experience and knowledge, she is also particularly interested in the pivotal role institutions have to play in enabling and supporting sustainable development. She has contributed to reports and publications on housing, renewable energy, general sustainability issues and governance.

Saliem Fakir is the Head of the Policy and Futures Unit at the World Wide Fund for Nature, South Africa. The unit's work is focused on identifying ways to manage a transition to a low-carbon economy. Saliem Fakir previously worked for Lereko Energy (Pty) Ltd (2006), an investment company focusing on project development and financial arrangements for the renewable energy, biofuels, waste and water sectors. He served as Director of the International Union for Conservation of Nature and Natural Resources, South Africa (IUCN-SA) office for eight years (1998–2005). Prior to IUCN, he was Manager for the Natural Resources and Management Unit at the Land and Agriculture Policy Centre. Between 2002 and 2005, he served as Chair of the Board of the National Botanical Institute. He also served on the board of the Fair Trade in Tourism Initiative and was a member of the Technical Advisory Committee of the Global Reporting Initiative in Amsterdam. He currently serves on the advisory board of Inspired Evolution One – a private equity fund for clean technology. He is also a columnist for the South African Centre for Civil Society and *Engineering News*. Saliem's qualifications are: B.Sc. Honours in Molecular Biology (WITS), Masters in Environmental Science, Wye College, London, and a senior executive management course at Harvard University in 2000.

Marie Blanche Ting is currently a doctoral researcher based at the Science Policy Research Unit (SPRU), University of Sussex, UK. Prior to starting her Ph.D., she worked for four years at South Africa's Department of Science and Technology (DST) as a Senior Specialist: Bio-economy, and before that she worked for more than four years at Sasol as a Senior Scientist with a special focus on environmental technologies. She holds a Masters in Climate Change and Development from the Institute of Development Studies (IDS), UK as a Mandela-Sussex Scholar. She also holds a Masters in Bioprocess Engineering from UCT. Her interests are in the intersection between science, technology, policy, and sustainable development.

Fumani Mthembi is the founding member of the Pele Energy Group, an energy utility research and development organisation. She is responsible for Knowledge Pele, the research and development advisory of the Pele Group. Her clients range from universities to independent power producers and the thematic focus of her work ranges from social assessments to mapping informal economies and risk and trial experiments. Her primary function is to set the research agenda and manage the business. Fumani has a BA Honours in Politics and Development Studies, Wits University and a Masters in Science, Society and Development, Sussex University.

Professor Lynn Krieger Mytelka is former Director of UNU-INTECH, the Netherlands, and Director of the Division on Investment, Technology and Enterprise Development, UNCTAD, Geneva. She taught at Carleton University, Ottawa, Canada, where she was Professor, Faculty of Management and Public Policy and subsequently a Distinguished Research Professor. She held an Honorary Professorship at the University of Maastricht, the Netherlands until 2010. Her research covers a broad range of issues in development; science and technology, including innovation systems; clustering and technological upgrading in traditional industries; North-South cooperation; biotechnology; strategic partnerships; multinational corporations, and competitiveness. Currently she is on the Executive Committee of the Global Energy Assessment, and has undertaken numerous research and capacity-building projects and consultancies for various governments and international agencies, including the EU, OECD, UNDP, UNCTAD, UNIDO, IDRC and the World Bank. Her recent publications are *Making Choices About Hydrogen: Transport Issues for Developing Countries* (edited with Grant Boyle, 2008) and *Innovation and*

Economic Development (ed.), the International Library of Critical Writings in Economics, Cheltenham, UK: Edward Elgar, 2007.

Simone Haysom is an independent researcher. She has several years' experience working for the think tank of the Overseas Development Institute in London, and has run her own consultancy, Cities in Flight, since 2013. She specialises in issues related to rapid urbanisation, migration, displacement, and humanitarian assistance. She has an M.Phil. from the University of Cambridge, and is currently a Miles Morland Fellow.

T. J. Pilusa holds a Ph.D. in Mechanical Engineering from the University of Johannesburg. He has more than eight years' experience in the mining, metallurgy and waste management industries. He is a member of the South African Institute of Mechanical Engineers (SAIMechE). He has published more than 35 international peer reviewed and refereed scientific papers in journals and conferences. He is a recipient of several awards and scholarships for academic excellence. His main areas of research are in alternative fuels, waste to energy, environmental pollution and waste management. His research involves classifications of industrial wastes, energy recovery, beneficiations processes and energy utilisation mechanisms.

Edison Muzenda is a Full Professor and Head of the Chemical, Materials and Metallurgical Engineering Department at Botswana International University of Science and Technology. He was previously a Full Professor of Chemical Engineering, the Research and Postgraduate Coordinator, as well as Head of the Environmental and Process Systems Engineering and Bioenergy Research Groups at the University of Johannesburg. Professor Muzenda holds a Ph.D. in Chemical Engineering from the University of Birmingham, United Kingdom. He has more than 16 years' experience in academia in institutions such as the National University of Science and Technology, Zimbabwe; University of Birmingham; University of the Witwatersrand and the University of Johannesburg. His current research activities are mainly focused on waste-to-energy projects with the South African National Energy Development Institute (SANEDI), the City of Johannesburg (Pikitup) and the Recycling and Economic Development Initiative of South Africa (REDISA). He has published more than 260 international peer reviewed and refereed scientific articles in the form of journals, conferences, books and book chapters, and supervised more than 30 postgraduate students.

Lyndall (Lynda) Mujakachi has experience in policy development, environmental management, local economic development and disaster management and has worked in the SADC region. Her research interests include sustainable energy in the built environment; the political economy of transitioning to a low-carbon economy; and knowledge exchange between universities, research institutions and government, focusing on adoption and adaptation of research knowledge and the requisite policy development. She has her own consultancy, Andly Consulting, and is a member of the South African Planning Institute (SAPI). She has a BSc. in Public Administration and MSc. in Rural and Urban Planning.

Dr Ogundiran Soumonni is a Senior Lecturer in Innovation Studies at the Wits Business School in Johannesburg, and a South African Research Chairs Initiative (SARChi) Research Associate at the Tshwane University of Technology. He obtained his Ph.D. in Public Policy from the Georgia Institute of Technology in Atlanta, Georgia, USA, where he focused on energy and environmental policy as well as innovation studies. Dr Soumonni's teaching interests include subject matter on creativity and innovation, 'technopreneurship', and philosophical paradigms in scientific research. His primary research interest lies in the area of innovation for sustainability from both a policy and a firm-level perspective. Some of his previous publications span subject matter on electricity policy, biofuels policy, nanotechnology policy, and innovation policy and management. Prior to embarking on his doctoral studies, he worked as a materials engineer in the area of energy-efficient lighting, following undergraduate degrees in physics and mathematics, and a Master's degree in materials science and engineering, respectively. He is an active member of the International Network on Appropriate Technology.

List of Abbreviations

In the List of Contributors

DST	Department of Science and Technology
FOCAC	Forum on China/Africa Cooperation
IDS	Institute of Development Studies
IERI	Institute of Economic Research
IISD	International Institute for Sustainable Development
IUCN	International Union for Conservation of Nature and Natural Resources
MPRDA	Mineral and Petroleum Resources Development Act
REDISA	Recycling and Economic Development Initiative of South Africa
S&T&I	Science, Technology and Innovation Policies
SAIMechE	South African Institute of Mechanical Engineers
SANEDI	South African National Energy Development Institute
SAPI	South African Planning Institute
SARChi	South African Research Chairs Initiative
SPRU	Science Policy Resource Unit
TERI	The Energy and Resources Institute
UNU-MERIT	United Nations University Maastricht Economic and Social Research Institute on Innovation and Technology
WWF	World Wide Fund for Nature

Chapter 1:	**The Trojan Horses of Global Environmental and Social Politics**
BASIC	Brazil, South Africa, India and China
CCS	Carbon Capture and Storage
COP	Conference of Parties
EST	Environmentally Sound Technologies
GATT	General Agreement on Tariffs and Trade
GDP	Gross Domestic Product
GHG	Greenhouse Gases
IAC	Inter-Agency Committee
IPCC	Intergovernmental Panel on Climate Change
IPR	Intellectual Property Rights
OECD	Organisation for Economic Co-operation and Development
PES	Payment for Environmental Services
RED	Radical Ecological Democracy
UN	United Nations
UNEP	United Nations Environment Programme
UNFCC	United Nations Framework Convention on Climate Change
US	United States
WB	World Bank
WCED	World Commission on Environment and Development

Chapter 2:	**LTMS and Environmental and Energy Policy Planning in South Africa: Betwixt Utopia and Dystopia**
BAU	Business as Usual
CCS	Carbon Capture and Storage
DERO	Desired Emission Reduction Outcomes
GWC	Growth Without Constraints
IPPC	Intergovernmental Panel on Climate Change
IRP	Integrated Resource Plan
LTMS	Long-term Mitigation Scenarios
MEC	Minerals-Energy Complex
MPA	Mitigation Potential Analysis
MW	Megawatts
NAMAs	Nationally Appropriate Mitigation Actions
NDP	National Development Plan
PICC	Presidential Infrastructure Coordinating Commission
PPD	Peak, Plateau and Decline Projections
R&D	Research and Development
RBS	Required By Science
SIPS	Strategic Integrated Projects
WHR	Waste Heat Recovery

Chapter 3:	**Historical Review of the Relationship between Energy, Mining and the South African Economy**
ANC	African National Congress
BEE	Black Economic Empowerment
COSATU	Congress of South African Trade Unions
CSP	Concentrated Solar Power
CTL	Coal to Liquids
DEAT	Department of Environmental Affairs and Tourism
DME	Department of Minerals and Energy
DMR	Department of Mineral Resources
DOE	Department of Energy
DTI	Department of Trade and Industry
EIUG	Energy Intensive Users Group
GHG	Greenhouse Gases
GW	Gigawatt
GWh	Gigawatt hours
IEA	International Energy Association
IEC	Independent Electoral Commission
IPP	Independent Power Producers
IRP	Integrated Resource Plan
ISMO	Independent Systems Market Operator
LTMS	Long-term Mitigation Scenarios
Mt	Million tons
MPRDA	Mineral Petroleum Resources Development Act

MEUs	Municipal Electricity Undertakings
NDP	National Development Plan
NEDLAC	National Economic Development and Labour Council
NEMA	National Environmental Management Act
NUM	National Union of Mineworkers
NUMSA	National Union of Metal Workers of South Africa
OECD	Organisation for Economic Co-operation and Development
PGMs	Platinum Group Metals
PPD	Peak Plateau Decline
PV	Photo Voltaic
SACP	South African Communist Party
SADC	Southern African Development Community
SIMS	State Intervention in the Minerals Sector
Tpc	Tons per capita
Twh	Terrawatt hours
Chapter 4:	**Lost in Procurement: An Assessment of the Development Impact of the Renewable Energy Procurement Programme**
BBBEE	Broad-based Black Economic Empowerment
CLO	Community Liaison Officer
COD	Commercial Operations Date
CSI	Corporate Social Investment
DFI	Development Finance Institution
DOE	Department of Energy
ED	Economic Development (Obligations)
EPC	Engineering, Procurement and Construction (Contractor)
IDP	Integrated Development Plan
IPP	Independent Power Producer
M&E	Monitoring and Evaluation
NDP	National Development Plan
NERSA	National Energy Regulator of South Africa
O&M	Operations and Maintenance
PPA	Power Purchase Agreement
PRA	Participatory Rapid Appraisal
RE	Renewable Energy
REIPPP	Renewable Energy Independent Power Producer Procurement Programme
SED	Socio-economic Development
Chapter 5:	**Making Transitions to Clean and Sustainable Energy in the South African Urban Transport Sector: Linkages to Growth and Inclusive Development**
BRT	Bus Rapid Transit
CCI	Clinton Climate Initiative
DTBSA	Daimler Truck and Bus Groups in South Africa

EU	European Union
HFCs	Hydrogen Fuel Cells
HFCV	Hydrogen Fuel Cell Vehicle
ITP	Integrated Transport Plan
MBSA	Mercedes Benz South Africa
NECAR	New Electric Car
OFMSW	Organic Fraction of Municipal Solid Waste
PEM	Proton Exchange Membrane
SACN	South Africa Cities Network

Chapter 6: **'Green' policymaking and implementation at city-level: lessons from efforts to promote commuter cycling in Johannesburg**

BRT	Bus Rapid Transit
C40	C40 Cities Climate Leadership Group
CBD	Central Business Districts
DA	Democratic Alliance
GCRO	Gauteng City Region Observatory
ICLEI	Local Governments for Sustainability
JUCA	Johannesburg Urban Cyclists Association
KWV	German Bank for Reconstruction and Development
NMT	Non-Motorised Transport

Chapter 7: **The Energy and Water Nexus: The Case for an Integrated Approach for the Green Economy in South Africa**

AMD	Acid Mine Drainage
CCS	Carbon Capture and Storage
CSP	Concentrated Solar Power
EWN	Energy and Water Nexus
FGD	Flue-gas Desulfurisation
GTL	Gas-to-Liquids
GW	Gigawatts
IRP	Integrated Resource Plan
MDG	Millennium Development Goals
R&D	Research and Development
RE	Renewable Energy
Tcf	Trillion cubic feet
WMA	Water Management Areas

Chapter 8: **Waste Re-Use: Oil Extraction from Waste Tyres and Improvement of the Waste Tyre Industry**

ASTM	American Society for Testing and Materials
CAPEX	Capital Expenditure
CHFO	Crude Heavy Fuel Oil
CI	Compression Ignition

CIE	Compression Ignition Engine
CoJ	City of Johannesburg
CoT	City of Tshwane
CTDO	Crude Tyre Derived Oil
DF	Diesel Fuel
DoE	Department of Energy
EBITDA	Earnings Before Taxes, Depreciation and Amortisation
EIA	Environmental Impact Assessment
Ekh	Ekurhuleni
ESD	Electrostatic Discharge
ETRMA	European Tyre and Rubber Manufacturers Association
GDP	Gross Domestic Product
HFO	Heavy Fuel Oil
ICE	Internal Combustion Engine
IWTMP	Integrated Waste Tyre Management Plan
MISTRA	Mapungubwe Institute for Strategic Reflection
NDP	National Development Plan
NOPAT	Net Operating Profit After Tax
NRF	National Research Foundation
OPEX	Operating Capital
PBIT	Profit Before Interests and Tax
REDISA	Recycling and Economic Development Initiative South Africa
ROD	Record Of Decision
RSA	Republic of South Africa
SANEDI	South African National Energy Development Institute
TDF	Tyre Derived Fuel
TPA	Tons Per Annum
US$	United States Dollar
UV	Ultraviolet
ZAR	South African Rand

Chapter 9: Energy Efficient Low-income Housing Development in South Africa: The Next Build Programme

ANC	African National Congress
DoE	Department of Energy
FBAE	Free Basic Alternative Energy
FBE	Free Basic Electricity
GWh	Gigawatt hour
IDP	Integrated Development Plan
iEEECO	Integrated Energy Environment Empowerment and Cost Optimisation
kWh	Kilowatt hour
MFMA	Municipal Finance Management Act
MW	Megawatt
PV	Photovoltaic

R&D	Research and Development
RDP	Reconstruction and Development Programme
REIPPP	Renewable Energy Independent Power Producer Procurement Programme
SANS	South African National Standards
SHS	Solar Home System
SIP	Strategic Integrated Project
SME	Small and Medium Enterprise
SWH	Solar Water Heater

Chapter 10: Off-grid Renewable Electrification as a Viable and Complementary Power Planning Paradigm in Southern Africa: A Quantitative Assessment

AC	Alternating Current
ADP	Abiotic Depletion Potential
AP	Acidification Potential
BE	Basic Electricity
CCGT	Combined Cycle Gas Turbine
CF	Capacity Factor
CHP	Combined Heat and Power
COE	Cost of Electricity
CRF	Capital Recovery Factor
DC	Direct Current
DG	Distributed Generation
DoE	Department of Energy
DRC	Democratic Republic of Congo
EIA	Environmental Impact Assessment
EP	Eutrophication Potential
FBAE	Free Basic Alternative Energy
FBE	Free Basic Electricity
GHG	Greenhouse Gas
GWP	Global Warming Potential
HOMER	Hybrid Optimisation Model for Electric Renewables
IEA	International Energy Agency
INEP	Integrated National Electrification Programme
LCA	Life Cycle Assessment
LCIA	Life Cycle Impact Assessment
LCOE	Levelised Cost of Electricity
NASA	National Aeronautics and Space Administration
NPC	Net Present Cost
NREL	National Renewable Energy Laboratory
NuRA	Nuon-RAPS
PED	Primary Energy Demand
POCP	Petrochemical Ozone Creation Potential
PPP	Purchasing Power Parity

PV	Photovoltaic
RET	Renewable Energy Technology
SADC	Southern African Development Community
SAPP	Southern African Power Pool
SHS	Solar Home System
US EPA	United States Environmental Protection Agency
WAPP	West African Power Pool

Introduction

Radhika Perrot

Over the past couple of years, South Africa adopted various policy measures to promote the development of renewable energy and other low-carbon and sustainable technologies, mainly driven by a concern to reduce emissions by more than 30 per cent by the beginning of the next decade and to make low-carbon sources of energy a substantial part of the total energy mix. Although the goal is ambitious, insufficient progress has been made in terms of development and deployment of low-carbon and environmentally improved technologies and innovations. South Africa is still the most energy-intensive country per capita (as compared to other BASIC countries, i.e. Brazil, India and China), and national energy consumption data for 2006 reveals that 95 per cent of the country's electricity is generated from coal and 75 per cent of the country's primary energy comes from coal.

Although efforts have been made to include renewable energy in the national total energy mix[1] through the Renewable Energy Independent Power Producers Procurement Programme (REIPPP), there are criticisms that the amount allocated to renewable energy generation capacity until 2030 (3,725 MW) is not enough to make a substantial contribution to the total energy mix and for the programme to be eventually effective in generating growth through new industries and jobs. The nuclear power component, on which much of the hoped-for reduction in carbon emissions seems to swing, is a controversial option that would, in any case, not have come on stream by 2020. Moreover, the plausibility of the energy mix proposed in the Integrated Resource Plan (IRP 2010–2030) to reduce emissions has not been adequately tested, and the validity of its assumptions on projected demand has already been questioned by a draft revision of the document.

The MISTRA research project on which this book is based set out to investigate and understand the low-carbon path undertaken by South Africa in its drive to mitigate emissions and make low-carbon sources of energy a substantial part of the total energy mix. Various government departments are engaged in relevant sectors of the economy, adopting a myriad policies and

1. The Integrated Resource Plan of 2010–2030 outlines the required new generation capacity for the next 20 years, and it takes into account the different energy carriers required to ensure that stability for the national energy supply and renewable energy are part of this mix.

strategies in an effort to mitigate greenhouse gas (GHG) emissions. But often such efforts are constrained by factors that characterise the transition to low-carbon systems, namely the path-dependent nature of South Africa's institutions, industry and large organisational incumbents.

Fossil-fuel-based electricity generation and transportation systems have together created techno-institutional complexes in South Africa that have *locked-out* the development of new sustainable technologies, including a transition to a low-carbon economy based on renewable energy technologies. Technological and institutional *lock-ins* act as barriers to sustainable innovation and they occur through combined interactions among socio-economic and socio-technological systems and governing institutions. In fact, such dependencies disable governing institutions from fully comprehending and effectively implementing new technologies, ideas and knowledge because new and sustainable innovations require new sets of institutional capabilities, knowledge bases and, most importantly, an integrated approach towards policymaking.

Issues related to the transition to a green economy and sustainable development such as path dependencies and carbon lock-in demand a revision[2] of both development processes and institutions and organisations that have been built on old structures, behaviours and systems. The embeddedness and ubiquity of the current system in South Africa suggests the need for conventional policy and decision-making in silos to give way to an integrated approach that reduces trade-offs and builds synergies across energy systems and interrelated sectors.

In an effort to understand the South African green economy landscape better, the various chapters of this book examine issues ranging from global and domestic climate change and sustainable energy issues, to the mineral-energy complex issues that have given rise to local and sector-specific problems. Each chapter seeks to convey policy choices and recommendations, but at the centre of all these recommendations is a clear articulation of the need for an integrated mix of policy instruments in South Africa to mitigate emissions and promote the development of a low-carbon economy through the low-carbon and sustainable energy technologies and low-carbon innovation across various sectors of the economy.

Three different but overlapping major themes are covered in the book:

Section I analyses the *co-evolutionary role of governments, civil societies*

2. A quote often attributed to Albert Einstein, i.e. 'You cannot solve a problem using the same thought process that created it', is appropriate in this regard.

and industries and the influence of their interactions on sustainability actions and on climate change mitigation strategies in South Africa.

Chapter 1 begins with a critique of the global discourse on sustainable development and its recent reincarnation since the 1992 Earth Summit in Rio – the green economy. The evolution of the discourse on global warming, the courses of action and nature of negotiation that evolved, including an understanding of how these have been articulated and interpreted by countries in the North and the South are discussed. The chapter provides a few considerations of the critical need to reframe current discussions around sustainable development and the green economy through an understanding of interpretative flexibilities from a sociological perspective; and of the need to recognise the interdependencies of sustainable development such as society, the economy, technology and the natural environment.

Chapter 2 brings the discussion to South Africa where it critically analyses the climate change discourse within the Long-term Mitigation Strategy (LTMS) process, and points out challenges of environmental and energy planning within its context. The chapter argues that energy and environmental policies in the country have been hampered by policy dissonance because they largely ignore 'data from the real world' such as vested interests, the political economy, and the critical role of coal, which is still a key driver of the economy. In South Africa, a cheap and abundant supply of coal is both a determinant of existing energy policies and an obstacle in the transition to a low-carbon economy. Although the LTMS provides a good overview of the climate change mitigation challenge in South Africa, there are a number of reasons as to why the LTMS is not working, which the chapter sets out to analyse. Further, climate policies of seven countries have been analysed for a conceptual understanding of what constitutes a 'coherent' or integrated climate policy for a comparative understanding of the country-specific mitigation challenges and the effect of fossil fuel path-dependencies on the transition to a low-carbon economy in each country.

The major challenge for sustainable development in South Africa or any other country is in the energy sector, and in South Africa it accounts for more than 70 per cent of greenhouse gas (GHG) emissions. In South Africa, the strong influence of the power relations in the energy sector has been observed and discussed in Chapter 3. The chapter analyses the feasibility of sustainable energy development in the mining sector which has co-evolved and been structured around interests, ideas, and institutions that have locked in its path to a high-carbon intensive economy. The influence of power and lobbying by groups such as mining businesses and trade unions is examined here. The technological and institutional lock-ins that characterise the South Africa economy have sustained a minerals-energy complex that impacts not only on climate change issues, but also the broader challenges of poverty and inequality.

Chapter 4 examines the socio-economic development limitations of the renewable energy programme in South Africa – the Renewable Energy Independent Power Producers Procurement Programme (REIPPP). It analyses the design of the renewable energy on-grid procurement system that sets out development socio-economic targets in rural and poor communities in South Africa. The chapter then makes recommendations on incorporating an expanded interpretation of development that would ensure that the government is using the right indicators to measure the impact of the programme on development.

Section II of the book analyses the dynamics in the *transition to a low-carbon economy* through an exploration of a number of sectors such as transportation, water usage and waste. Such an analysis is intended to improve an understanding of the processes involved in the formation and uptake of sustainable technologies, thereby identifying the key associated challenges for policymakers that are managing the transformation process. Local and city-level authorities have become important institutional players for policy decisions around sustainability and pollution, all of which have considerable impact on climate change. There is, therefore, a need to identify the opportunities at these local or city-level spaces, including, in particular, the transportation and waste sector, which accounts for a significant proportion of GHG emissions in the country.

As a part of the process towards this understanding, Chapter 5 analyses and compares the development of the bus rapid transit (BRT) systems in Johannesburg and Cape Town and the extent to which their development has played a role in stimulating a move towards energy transitions to low-carbon technologies. The chapter analyses the cases of the Rea Vaya in Johannesburg and MyCiTi in Cape Town, while investigating past policy practices, norms and habits that have created institutional and technological lock-ins, and provides recommendations on how such lock-ins and dependencies can be overcome. The development of the BRT systems is largely in the hands of provincial governments and major cities that make their own choices (often different from each other) about the purchase of fuels and bus type. Within the country, and between the two cities analysed, the development of multiple, new and competing types of buses has given rise to uncertainties in choices that might not survive a transition to clean fuels in South Africa if it continues on the current 'development' path. This is likely to lead towards an incongruent path, leading to a multiplicity in the development of know-how and skills in the local bus manufacturing capacity within the country, ultimately to duplication of efforts and waste of resources.

As a city-level case exploration, Chapter 6 analyses the efforts to promote the 'green' goal by the City of Johannesburg, whilst analysing the practices and changes that have created the current momentum. Its findings point to the critical role played by some actors who are from outside the domain of city-level governance efforts such as cycling advocacy lobbies and universities. This chapter and the chapters in this section show that the transition to low-carbon energy systems in transportation will require new and paradigmatic changes at the institutional, technological and consumer levels.

The two biggest challenges facing South Africa today are the frequent electricity outages and the potential water crisis. Whilst the electricity crisis is receiving the required attention for an appropriate and timely intervention, little is discussed about an impending water crisis that is waiting to explode in an extremely water-scarce country such as South Africa. The biggest cause of both the crises is ironically not from increased demand for energy or water usage but rather from insufficient maintenance of water and energy infrastructure, including the negligence of energy and water conservation policies and practices in the country.

Chapter 7 explores the relationship between energy and water and elaborates on the critical need for a parallel and integrated development of water and energy policies in South Africa. Often in policy debates, the connection between water and energy is never well articulated while evidence suggests a strong codependence of electricity production from coal and water use, and how a shortage of water will affect electricity availability. Unfortunately, even future planning documents and tools such as the IRP 2030 do not consider the risks of potential water scarcity for the planned generation capacity and electricity supply or the electricity sector's ability to provide reliable and sustainable energy supply in the event of water insecurity.

Changes in the production processes of industries, institutional changes and consumer habits that are required in the transition to low-carbon energy systems often take considerable time. In the shorter term, however, increases in energy efficiencies such as the conserving of energy in buildings, industrial processes and co-generation of energy in industries are good and successful examples of energy saving and conservation practices, where benefits such as reducing GHG emissions are achieved without increasing energy cost through new or additional energy generation capacity.

Evaluating a category of energy-efficient and energy-conservation measures and practices in South Africa, Chapter 8 analyses the particular case of the waste tyre industry in the province of Gauteng, and the associated policy regulation and industry practices along the value-chain of its waste tyre disposal. Approximately 55 per cent of the 7.25 million new tyres produced in 2011 were sold in Gauteng, the majority of which would end up as waste and add to the over 70 million tyres that have so far been piling up as scrap tyre waste. The Department of Environmental Affairs (DEA), through the Recycling and Economic Development Initiative of South Africa (REDISA), has recently instituted a R2.30 per kilogram levy on all tyre manufacturers to ensure that tyres do not end up on waste dumps. And by evaluating the best method in reusing or recycling waste tyres, the chapter provides a quantitative assessment of the economic feasibility of pyrolysis as a method for waste-tyre treatment, including generation of energy from the wasted tyres. It also provides recommendations on how to make waste tyre treatment methods such as pyrolysis economically feasible including the creation of economic opportunities such as new jobs.

After the end of apartheid in 1994, a number of policies and programmes were initiated that set out to reconstruct and assist in achieving the country's social and developmental goals. One such initiative was the development of housing for the low-income population, called the Reconstruction and Development Programme (RDP). Chapter 9 evaluates the RDP programme in terms of gauging the implementation of energy efficiency in these households, including the implementation of renewable energy technologies and solutions such as rooftop solar. The chapter highlights the need and importance of the coordination of policy action between national and provincial government and departments, and continual monitoring and evaluation of these projects post implementation.

Section III of the book *explores regional optimal energy strategies* as there is compelling evidence from many countries in Africa that shows the need for off-grid renewable energy systems, and the importance and benefits of which are gaining ground on the continent. Policymakers need to send strong signals to all development and regional partners within the Southern African Development Community (SADC) of their commitment to the development of renewable energy resources as part of the process of sustainable development within the continent. Chapter 10 makes a quantitative assessment of the technological, economic and environmental implications of extending electrification through off-grid electrification solutions in southern Africa. The chapter demonstrates how solar photovoltaics and wind turbines can meet the basic electricity requirements of individual households in non-electrified areas and provides policy implications for increased energy access, sustainability and technological competence building up in the region.

Section IV summarises in point form the policy implications and individual recommendations arising from all the chapters. Overall, this book argues that the complex and interrelated nature of the transition to a low-carbon economy, and towards sustainable development, demands that the issues and concerns pertaining to the green economy be resolved through a deeper understanding of the interrelated nature of these issues, and it is thus important that policy approaches are devised that are integrated in their approach.

Policymaking processes must attempt to take cognisance of the socio-technical practices and institutional factors that contribute to sustainable development and thereby build synergies across industrial and energy systems and interrelated sectors.

SECTION I

Co-evolutionary Role of Governments, Civil Societies and Industries

CHAPTER 1

The Trojan Horses of Global Environmental and Social Politics

A Discourse on Sustainable Development, the Green Economy and Climate Change

Radhika Perrot

One has to make up his mind whether he wants simple answers to his questions – or useful ones ... You cannot have both.
J. A. Schumpeter

ABSTRACT

This chapter offers a critique of the dominant discourse on sustainable development, the green economy, and climate change policies from the perspective of developing countries. It examines how the hegemonic discourse around market-led economic growth principles, such as green growth, have led to decades of mitigation deadlocks and further widened the gulf between the countries of the North and the South. Indeed, a few considerations are critical in reframing the discussions around sustainable development and its recent reincarnation – the green economy – through an understanding of interpretative flexibilities from a sociological perspective; and of the interdependencies of sustainable development such as society, the economy, technology and the natural environment.

THE OXYMORON OF SUSTAINABLE DEVELOPMENT

Our Common Future[3] was the first document that placed environmental issues firmly on the political agenda and made it an international mandate. It aimed to discuss the environment and development as one single issue with focal attention given to population, food security, the loss of species and genetic resources, energy, industry, and human settlements. The report sought to address these critical issues by proposing new forms of international cooperation whilst a crucial part of its mandate was to 'raise the level of understanding and commitment to action on the part of individuals, voluntary organizations, businesses, institutes, and governments' (1987: 347). An international scientific committee was constituted in 1988 called the Intergovernmental Panel on Climate Change (IPCC),[4] which aimed to raise the level of understanding of human-induced climate change through scientific information and evidence.

Although the Report understood the complexity and interdependence of the impact of human-induced climate change (such as industries, land, forestry, population and sub-sectors), it is the lack of analysis around economic growth and its sustaining modes of production and consumption that diluted an in-depth understanding and discussion of a development that was truly sustainable. Many developing countries not only had cheap resources and labour, they had less stringent environmental regulations than developed countries. As a result, many large industrialised countries set up pollutive factories and other industries in many developing countries of the world that had low environmental standards or weak enforcement.[5] Extractive industries such as coal, metals and petroleum are major sources of environmental damage with a direct exploitative relationship with the natural environment, and yet, the effects of pollution from such industries had received less attention (Stollery, 1985). So, in the absence of a critical analysis of economic growth and the market-led principles governing it, the report further postulated that such growth could be reformed and expanded (Ahmed, 2004) and unsuccessfully attempted to bring together environmental protection and economic expansion (Hove, 2004).

3. Also known as the *Brundtland Report*, it was published by the United Nations World Commission on Environment and Development (WCED) in 1987.

4. The IPCC assesses the scientific, technical and socio-economic information relevant for the understanding of the risk of human-induced climate change. The IPCC produces reports that support the United Nations Framework Convention on Climate Change (UNFCCC), which is the main international treaty on climate change. It does not carry out its own original research, nor does it do the work of monitoring climate or related phenomena itself, but it bases its assessment on the published literature, which includes peer-reviewed and non-peer-reviewed sources (IAC, 2010).

5. The pollution haven hypothesis posits that, when large industrialised nations seek to set up factories abroad to avoid the cost of meeting the high environmental standards in their own country, they will often look for the cheapest option in terms of resources and labour that offers the land and material access they require. However, this often comes at the cost of environmentally sound practices.

Section III, Article 27 of *Our Common Future* postulates that:

> *In the short run, for most developing countries except the largest a new era of economic growth hinges on effective and coordinated economic management among major industrial countries – designed to facilitate expansion, to reduce real interest rates, and to halt the slide to protectionism. In the longer term, major changes are also required to make consumption and production patterns sustainable in a context of higher global growth.*

The changes suggested, regarding the modes of production and consumption of existing economic growth trajectories, which were in effect responsible for the degradation of the environment, were not identified; nor were the modes in those of developed countries seriously questioned. Rather, the 'changes' and 'expansion' for sustainable development were suggested for developing countries, with explicitly guided considerations of what was perceived to be 'higher economic growth'.

Soon thereafter, in 1991, a Hague Report on Sustainable Development published the outcomes of a symposium to inform the upcoming UN 1992 Earth Summit in Rio. This report, however, acknowledged that it was impossible to pursue a global dualistic model of development – one which suggested that the North would continue to pursue the same path to material consumption, while another model was suggested for the poor South. The report estimated that, 'If the same material standards had to be replicated in the South, it would require 10 times the present amount of fossil fuel and roughly two hundred times as much mineral wealth' (*The Hague Report*, 1991).

Thus began the *junkie politics* of sustainable development and climate change[6] between the countries of the North and the South. The sceptical countries of the South, with their colonial and neocolonial experiences imposed by the North, felt they were once again being handed down solutions and impositions, many in the form of restrictive trade agreements that undermined economic growth. One of the recommendations in the *Our Common Future* report, which had no direct evidence of a link to sustainable development, required developing countries 'to halt the slide to protectionism' (Article 27, Section III). The General Agreement on Tariffs and Trade (GATT), one of several major international agreements that sought to liberalise trade by removing protective tariffs, quotas, and other

6. One minute people ride high on NGO calls to save the planet, and in the next minute they lash out on the realities of international relations (Ruth, 2015). This, Ruth calls the junkie politics of the climate.

barriers, was established in 1994. Many formerly protectionist developing countries were slowly coerced into embracing free trade and market-led principles and joined the GATT, while around the same time, the United States and Europe, the original framers of the GATT, adopted protectionist measures on entire industrial sectors like steel, chemicals and electronics, including food and agriculture.

Although developed countries meekly acknowledged their role in massive environmental degradation and pollution in their quest towards industrialisation, they did not fully accept that high economic growth trajectories based on often harmful and long-cycle modes of consumption and production patterns had been the main cause. The infamous London smog of 1952, and the covering and diverting of the polluted river Zenne that runs through the city of Brussels in Belgium, filled with massive garbage and decayed organic matter, were cited by developing countries as examples of the developed countries' quest to industrialise without regard to air quality and environment.

So, amidst distrust between countries of the North and the South, it was therefore not surprising that in 1992 the Rio Earth Summit agenda on sustainable development failed to make a plausible impact. The notion of sustainability that emerged from this summit politicised the debate on environment and growth – negotiations between government, business, and 'pragmatic' environmentalists assumed that new markets and technologies could simultaneously boost economic growth and preserve the environment (Kallis, 2015). Environmental problems were largely confined to the realm of technological improvement (discussed below), and thus sustainability decisions and mitigation agreements were to remain within the domain of technical experts, technocrats and policy elites.

Production and consumption decisions and patterns that are directly linked to economic growth, and upon which most national growth policies are premised, were never questioned as the direct source of unsustainable practices. This outlook, which persists to this day, is especially problematic for countries that are still 'developing' or 'emerging', for the definition and intention of the goals of economic development (as always understood from the perspective of economic growth theory and practice) emerge as conflictual with sustainable development. The 1992 Earth Summit in Rio failed to provide a framework for action or to articulate specific and tangible ways in which the achievement of sustainable development could be put into action. In the process, the dire development situations of many developing countries were watered down (Hove, 2004). There was little appreciation of extractive production processes and wasteful consumption that sustain such production as part of the problem, which had led to rampant mining and deforestation.

So, since the 1992 Rio Summit, sustainable development has been lost in a myriad misinterpreted translations, and for many it 'appears at best an empty phrase and at worst a Trojan horse for the redefinition of the public interest by a powerful few' (Voß and Kemp, 2005, p. 3). It became a loosely interpreted word for Brazil, India and China (or the BASIC countries) that were emerging as economic strongholds in the early and late 1990s. In their context, social progress meant following the industrialised or developed countries in the pattern of building skyscrapers, bridges, concrete houses, and putting more cars on the road. Deforestation, uprooting indigenous people, and vanishing wetlands, hills and mountains were seen as unavoidable in the quest for such 'development' (Perrot and Soummoni, 2015). Modernisation and the imperative of high economic growth rates were pursued from the perspective of Modernisation Theory, which is largely based on the view that to develop means to become 'modern' by adopting Western cultural values and social institutions.

In all the discussions on sustainable development as articulated in *Agenda 21* – the summary document of the outcomes of the Rio Summit – developing countries were considered the main (and new) culprits of the environmental crisis. According to Lippert (2004), the three major causes of the environmental crisis (which formed the hegemonic discourse of sustainable development after the Summit) were:

- *Poverty in developing countries*

In their argument, poverty meant too little development and proposed that it could be overcome by economic growth or the trickle-down effect of economic growth. Such an argument also implied that countries of the North did not have to change their growth paths. Section 3.2 and 3.3 of *Agenda 21* explicitly argued: 'An effective strategy for tackling the problems of poverty, development and environment simultaneously [...is...] economic growth in developing countries that is both sustained and sustainable and direct action in eradicating poverty by strengthening employment and income-generating programmes.'

- *Population growth*

It was argued that if too many people depended on the limited resources of the planet it would create a stress on the earth's carrying capacity and destroy its ecological systems. Population stabilisation was proposed and population growth was seen as 'evidently' a problem of the developing countries. Fred Pearce, in his recent 2010 book *The Coming Population Crash: and Our Planet's Surprising Future*, disproved the half-century presumption that population growth is the driver of ecological apocalypse, arguing that it is not overpopulation that causes climate change it is overconsumption. Pearce (2014) writes, 'But why do we blame the poor in Africa for having babies when the real issue is overconsumption closer to home? It is the ravenous demands of the rich world that is enlarging the human footprint on our planet – pumping greenhouse gases into the air, polluting the oceans, trashing forests and the rest. Any further rise in numbers of poor people will barely figure in that.'

- *Lack of ecological modernisation*

An opinion argued that the cause of the environmental crisis is the lack of 'ecological modernisation'. Such a focus is techno-centric as solutions to the crisis are seen as existing only with technical experts and users, ignoring the fundamental issue of socio-economic transformation. Evidently, all developing countries lacked the technological sophistication and machinery and technical knowledge for 'ecological modernisation', and such machinery and technological expertise[7] were evidently with the countries of the North.

FRAMING THE SUSTAINABILITY PROBLEM

As evident from global discourse from the 1990s to the present, sustainable development has been understood as an end state rather than as a kind of problem framing. It cannot be determined once and for all but it is rather about the organisation of processes and not about particular outcomes (Voß and Kemp, 2005). There are many different ways in which a problem can be resolved and different strategies that can search for solutions to reach sustainable social and technological development paths. According to Voß and Kemp, 2005, the process towards sustainable development goals is reflexive when there is recognition of the complexities and uncertainties of

7. Lippert (2004) criticises the heavy emphasis put on efficiencies and the greening of research and technologies to attain sustainable development and such an approach points at solutions for problems that are not realised by people but only by rational scientists and specialists.

the natural reality of the problem. When the complexity of an issue is not reflexively addressed, the required impact and importance of interdependencies of sustainability (viz. the interdependencies between society, the economy, the technological and the natural environment) are missed. This is precisely because socio-economic, technological, and ecological elements are embedded within each sustainability problem.

So a key feature of sustainable development is the enormous complexity of systemic interactions and levels, for the process of sustainable transition also takes place through interactions between multi-level structures – the local, regional and global levels. Further, the interaction between these levels and the sub-systems of each level, and the interactions within each, add to the structural complexity of the issue (Voß and Kemp, 2005). And without a proper understanding of the systemic interactions between the various meta-levels structures, it would be impossible to transform an existing system into a sustainable one or provide solutions to the issue. Such a multi-level understanding takes a sociological perspective into account by analysing the process of alignment between the different levels and each level's sub-systems and social and technical elements.

So then, the framing of a sustainability problem should come from a perspective that links the sociological with an evolutionary analysis. A sociological perspective gives importance to interpretative flexibility, which means that different actors have different understandings of the same idea. If an idea is to be accepted there has to be wide and shared understanding. At the climate change negotiations, interpretative flexibility persists around the setting of sustainable goals and emission targets – what it means to be sustainable or to grow economically widely differs between countries. An evolutionary perspective, on the other hand, understands the process orientation of sustainable development. It characterises the transitions of, say, an existing fossil-fuel-based transport or electricity system to a cleaner or more sustainable system, through an understanding of the characteristics of a systemic transition, akin to a biological evolution: *variation* amongst the actors of a system or a population; *retention* of systemic characteristics from one generation to the next; and *selection* of better or superior characteristics that survived a competitive environment.

Voß and Kemp (2005) have identified three features that contribute to the complexity of sustainable development problems: heterogeneous interactions, uncertainty, and path dependency. Within each system are a number of heterogeneous actors that not only have interpretative flexibility about sustainable development and sustainability goals, but interact with each other through heterogeneous institutions,[8] social values,[9] knowledge base, user practices and strategies, among others. As a system is being transformed into a sustainable one, there are characteristics from an old system that are retained (the retention feature of an evolutionary process). When transitions are path dependent and characteristic of all systemic transitions, they are influenced, constrained and enabled by structures that have grown out of particular historical developments (Voß and Kemp (2005). And due to the complex dynamics of such transitions, sustainable development paths cannot be predicted with certainty. Therefore, sustainability as an orientation towards societal development does generate ambiguous goals – social conflicts are inherent in the concept and need to be carried out with it (Voß and Kemp, 2005).

At a broader level, sustainability needs to be pursued across the entire landscape and across many levels in both rural and urban areas. There are several elements of sustainable alternative paths in thousands of initiatives, resistance struggles and movements for social transformation around the world. Kothari believes that the 'radical ecological democracy' (RED)[10] is an emerging framework, and puts collectives and communities at the centre of governance and the economy. Radical environmentalists recognise that sociological ecology, with its focus on connecting humans with one another, is at odds with growth that separates and conquers (Kallis, 2015). The RED approach offers a systemic approach to social transformation, resting on political, economic, sociocultural and ecological pillars.

In Kothari (2014), the new vision of a sustainable *political pillar* is one of participatory democracy in which political boundaries become sensitive to ecological and cultural diversity. Grassroots and local initiatives through village and city councils and communities must become embedded within larger institutions of governance. There are many examples of cases that are attempting to combine localisation with the larger-scale national and provincial decision-making.

In India, around 72 riverine villages have formed the Arvari River Parliament, which meets regularly to make ecological, economic and social decisions; India's north-eastern state of Nagaland has enacted legislation

8. Institutions define the rules and norms and regulate how actors in a system should interact with each other.
9. People and actors in a system hold different values, and if they evaluate an idea or option they make different decisions, and thus the interpretative flexibility.
10. Kothari, A. 2014.

empowering villages with substantial decision-making, including control over the allocation of government funds for education, health and power. In Venezuela, communal councils used social, cultural and economic relations to define geographical boundaries.

Further, according to Kothari (2014), the vision of a new *economic pillar* rests on efforts to decentralise control over natural resources[11] and production chains through ownership by cooperatives. There are many examples of Indian villages that have revitalised local economies and reduced social inequalities. For example, Jharcraft of India has enhanced the livelihoods of 250,000 families by providing credit, technological assistance, marketing opportunities and recognition of cooperative-led production.

The sociocultural pillar of the RED is one that maintains and combines traditional knowledge and wisdom as cultural sources to generate responses and appropriate adaptive mechanisms to ecological (and climate) change, social and political changes and uncertainty. It also includes the dismantling and democratisation of knowledge and science from its long monopolisation by corporations and the state. In fact, indigenous people have been active in international conventions on creating indicators for sustainability, justice and other goals (Kothari, 2014).

RED ideas and analysis are similar to the so-called de-growth imperative, which involves a rethinking of the organisation of society. De-growth alternatives began to emerge as the formal economy fell into crisis, and activities include urban gardens, producer-consumer cooperatives, decentralised forms of renewable energy, alternative food networks, and so on. These activities are geared towards a more locally based economy; they have short production and consumption cycles; they do not have a built-in tendency to accumulate and expand; and they are less resource intensive than their counterparts in the formal economy (Kallis, 2015).

Within the UN climate change discussions (and as signatories) there are many poor countries, many of them African, who are yet to benefit from growth. And, to some extent, de-growth in the countries of the North can provide an ecological space for countries in the South. For example, a strong carbon cap for countries in the North and better terms of trade for countries in the South can help compensate past carbon and resource debt and redistribute wealth between countries of the North and the South (Kallis, 2015). Such rethinking – particularly RED and the de-growth imperative – requires the spread of certain core values of sustainable development that are

11. Such control rests on the principle of subsidiarity: the belief that those living closest to ecosystems and resources have the greatest stake in them and at least some of the essential knowledge for managing them (Kothari, 2014).

geared towards equity and social justice: to empower every person to be a part of decision-making and its holistic vision of human well-being encompassing physical, material, sociocultural, intellectual and spiritual dimensions, including the sharing of non-monetary wealth (Kothari, 2014; Kallis, 2015).

Elsewhere, Davis (2015) argues that shared interest and common good are vital for a renewed climate diplomacy, especially in leading towards Paris COP21 in 2015, and will constitute, not so much benefits that accrue to global GDP, but shared values such as a love of nature; respect for the history, identity and traditions of other nations and people; careful tending to children's inheritance, and solidarity with the world's poor. In this sense, de-growth in the North should provide a space for ecological alternatives and visions and the re-emergence of shared practices, such as the practice of human solidarity reflected in *ubuntu* in Africa.

Green Economy: A Slippery Slope

A recent reincarnation of sustainable development is the green economy. In 2008, amidst confusion and ambiguity on what constituted sustainable development and sustainability goals, coupled with the need to assuage perceived risks and fears from the global economic crisis, organisations such as the OECD, UNEP and the World Bank introduced the terminology, the *green economy*.

While the concept has its utility in terms of placing the economy at the centre of the green discussion – and vice versa – to resolve the current environmental crisis, it does have inherent damaging implications. Ironically, the very forces that caused the economic crisis, such as unbridled markets and capital, are being identified as the beacons of hope for the green economy (Death, 2014; Perrot and Soummoni, 2015). In other words, we cannot uncritically look at market-based instruments such as putting a price on carbon and trading, the emergence of profit-driven green companies, and payment for environmental services (PES). The extent to which mechanisms and principles of market-capitalism can be used to advance green technologies and the commoditising of nature, especially in developing countries, has to be critically interrogated.

As a result, countries have remained confined to a pillar approach to sustainability – in which the environment, society, technology and the economy are viewed separately. In January 2015, during his visit to India, Barack Obama urged the country to reduce dependence on fossil fuels, to which the Indian government responded with the retort that 'the country was not under any pressure to major cuts in emissions as it will not set targets that will undermine efforts to end poverty' (*The Guardian*, 2015). Under the new government in India, and in less than a year, the Modi government, strongly in favour of businesses and economic growth, is weakening India's existing environmental laws and thereby threatening the rights of those who rely on the forests for their livelihoods (Mahapatra, 2014).[12]

Such fallacies of unwarranted assumptions are frequently made in political statements – that economic growth, and now a 'green economy' (narrowly framed in the manner of neo-liberalism) will combat poverty, create jobs and reduce income inequality. But growth is only a condition and not a guarantee for poverty alleviation (Christoplos, 2014). Many countries in the global South fear that an accelerated phase-out of carbon-based energy will bring the end not only of 'development' but of any meaningful prospect for global economic justice (Athanasiou, 2014).

The apprehension on the part of the countries of the South is further reinforced by the conduct of the countries of the North, who have time and again shown inconsistencies when it came to setting and achieving emission reduction targets. Canada, for example, was active in the negotiations that led to the Kyoto Protocol in 1997 and the Liberal government signed the accord in 1997 and their parliament ratified it in 2002. But when the Conservative Party came to power in 2006, Canada began strongly to oppose the Kyoto Protocol and the imposition of binding targets unless mitigation targets were also imposed on countries such as China and India.[13] Canada has had increases in annual emissions over the last 15 years, largely fuelled by the fossil-fuel sector, transport, and most recently expansions of the oil sands. A group led by former United Nations Secretary-General Kofi Annan is condemning Canada as an international climate laggard that falls short even of impoverished Ethiopia in the effort to combat global warming (McCarthy, 2015).

The other climate laggard amongst the developed countries is Australia, which is the highest per-capita emitter of greenhouse gases among the OECD countries. In October 2014, the Australian Prime Minister declared that 'coal is good for humanity' while opening a coal mine. He added that

12. The current Indian government is threatening to dilute a number of important environmental protection legislations such as the Environment Protection Act, the Wildlife Protection Act, the Forest Conservation Act, among others (Mahapatra, 2014).

13. As Annex II countries, India and China, or the BASIC countries, were exempt from GHG reduction requirements under the terms of the Kyoto Protocol.

coal is vital for the future energy needs of the world and people should stop the 'demonisation of coal'. An African Progress Panel, co-chaired by Kofi Annan and former US Treasury Secretary, Robert Rubin, says Canada and Australia 'appear to have withdrawn entirely from constructive international engagement on climate' (McCarthy, 2015).

It is in the context, not only of the broader philosophical issues but also against the backdrop of these experiences, that RED and de-growth theorists argue that the issue is about imagining and enacting alternative visions to modern growth-based development.

The goal of higher economic growth rates in the context of how economic growth has been perpetuated through market-led principles is well-nigh impossible without harming the environment. So the dominant discourse of the green economy as the driver of economic growth and reduction of poverty that is prevalent in the BASIC countries evidently has its serious limitations from political and ecological viewpoints. [14]

Death (2014) proposes two lines of critique to South Africa's green growth emphasis, which can be extrapolated to the emphasis on green growth strategies in India and China at the moment:

- The overall articulation of the green economy concept is not very coherent, and provides little potential to drive sustained economic growth. Green economy holds little promise of becoming the organising concept for development in South Africa.
- Even if South Africa pursues a green economic path in a sustained and significant manner, there are worrying implications and contradictions, such as a relentless narrow focus on economic growth against the backdrop of high levels of inequality; commodification of nature (PES and bio-economy); and the intensification of industrial agriculture and genetic modification.

If the goal is to fight poverty and reduce income inequality, the focus must be on who are directly exposed to climate change impacts and environmental risks, and a shift from top-down to bottom-up policies is needed; but currently the green-growth discourse reflects a bias against the poor (Christoplos, 2014) and countries are narrowly focused on securing their own national interests and avoiding costly commitments to emission reductions or long-term funding for adaptation (Roos, 2010). Unless the

14. See Death (2014) for a detailed critique of the South Africa green economy discourse.

poor are meaningfully involved in decision-making, green growth policies will prove counterproductive.[15]

INTERNATIONAL GOVERNANCE SYSTEMS OF GLOBAL CLIMATE CHANGE

When *Our Common Future* was first introduced, there was an assumption that all countries would coordinate their activities and work to achieve the common goal, but decades of international climate mitigation negotiations and discussions, and ensuing disagreements on greenhouse gases (GHG) or emission reduction strategies, including the failure of the top-down approach of the Kyoto Protocol mechanism, have shown that, in addition to ideologies, policy approaches and structures, lack of coordinated action(s) among countries have stymied international consensual agreements on emission reductions.

Coordination failure of complementary activities is a form of market failure that originates from a lack of information and lack of trust between economic actors (Lütkenhorst, *et al.* 2014). Coordination failures at climate change negotiations for the past few decades have resulted from a lack of trust and an undue focus on market-led economic growth that has further engulfed the countries of the North and South. Referring to the reality of the Anthropocene era, Falk (2014) writes that transnational and global political action is still largely entrapped in old habits of thought, action and interaction, and the myopic time horizons of the sovereign priority of states are too short to adequately respond to the deepest challenge of the human future.

Falk proposes two future trajectories on which the long-term success of the transition to global and national sustainable practices, and effective international coordination of activities, will depend: (1) a revolutionary change in political consciousness and (2) government interest and diplomacy that facilitate the pursuit of human and global interests. The former is actor-orientated, i.e. achieving transition without changing the structure of the world order (Falk, 2014), and the latter is system-orientated, which requires changing economic structures, technologies and institutions (Falk, 2014; Lütkenhorst, *et al.* 2014). There, however, needs to be substantial convergence between these two trajectories if global climate change discourse and agreements are to become effective.

15. A pro-poor green growth strategy emphasises participation, transparency, accountability and non-discrimination, which are based on the Swedish government's concept of human-rights-based development (Christoplos, 2014).

In the context of developing countries, it will be critical to explore government failure and sustainable governance processes and structures that are not championing environmental causes. According to Lütkenhorst, *et al.* (2014), government failure is always part of the equation of political decision-making and may be as great a risk as, or even greater than, market failure. In democratic societies, it is expected that the predominant form of regulatory or political capture is derived from lobbying rather than straight corruption (Lütkenhorst, *et al.* 2014). But, in the absence of environmental lobbying, or interest groups' effectiveness in decision-making in countries such as South Africa, India and China, as prevalent in more developed democracies, it would be important to analyse what groups or whose economic and/or political interests climate change policies are serving.

Climate Change Response of the BASIC Countries

A 2014 special issue of the Oxford Development Studies explores how few and new 'rising powers' are currently challenging global 'rules of the game' on social and environmental issues. These rising powers are the emerging economies of Brazil, China, India and South Africa (BASIC), and are characterised by six distinctive criteria[16] that make them different from other technologically advanced strong growth economies such as South Korea and Taiwan (Nadvi, 2014). They will matter, and critically, in global discussions of climate change and in the setting of global environmental regulations and standards, for they hold immense potential to impact global environmental conditions, largely because of their increasing significance, not only on global production, but also on consumption. These countries have increasing interdependence with the rest of the world in terms of trade and capital flows, and thus have huge global environmental impact implications (Nadvi, 2014).

As the environmental Kuznets curve hypothesises,[17] and as these countries (the BASIC economies in our analysis) are expected to have better economic conditions, there is interest in knowing if domestic demands for better social and environmental quality and standards will improve once a certain level of development or growth has been achieved, as measured against how and what the industrialised countries have achieved. It is also critical to know, in

16. See Nadvi (2014) for a characterisation of BRICS: strong economic growth since the 1990s; significant participation in global trade; a large domestic market; strong state involvement in the economy; availability of private and public capital for investment; growing space for civil society in public-private discourse.

17. The environmental Kuznets curve hypothesises that in the early stages of economic growth, environmental degradation and pollution increase, but beyond a certain level of income per capita (which will vary for different indicators) the trend reverses, so that at high-income levels economic growth leads to environmental improvement. This implies that the environmental Kuznets curve is an inverted U-shaped function of income per capita.

their quest for economic development and growth, if these countries will attach importance to the ethical aspects of their production and consumption decisions or economic growth. Some of these countries, such as India and China, have already extensively and irreversibly compromised on domestic social and environmental quality and standards in favour of rapid and unsustainable economic growth.

The interpretative flexibility of how countries have approached environmental and sustainable issues thus far, and questions concerning trade-offs between economic growth and environmental quality and standards, have critical implications for how global negotiations around climate change will be directed. This applies to future post-Kyoto climate discussions, including the upcoming COP 21 in Paris in December 2015. The BASIC[18] voices, or the rising powers, will determine if any international agreements will be reached. Currently, the 'ethical' debate of these countries has tended towards trading off the environment for 'development': sustainable development has been given little bargaining space, as most often myopic visions of leaders, a lack of political will and 'blame-games' have constrained international greenhouse gas (GHG) mitigation and adaptation agreements and strategies. Social and environmental quality and (sustainable) economic growth are looked upon as trade-offs and mutually exclusive and this is because environmentalists, business leaders and policymakers have all made the environment an economic rather than a political issue (Unmüβig, 2014).

One the biggest challenges for the BASIC countries remains – one of initiating and implementing forms of ecological modernisation and pathways that will likely lead towards sustainable growth and development. The BASIC countries should strive to re-frame the climate change negotiations towards discussion of sustainability solutions, exchange experiences and give moral leadership to the global negotiations by positing approaches that transcend the current paradigms. In this way, they will help unlock the decades-long North-South logjam around global sustainability, climate change and development, primarily by winning over the majority of humanity across the globe towards approaches that are reasonable, just and socially sustainable.

18. This analysis, however, will refer to the BASIC countries as were constituted as a negotiation group for the COP in Copenhagen in 2009 and signed the Copenhagen Accord.

Figure 1: Mitigation policies are enshrouded in market-led principles driving economic-growth agendas that cannot resolve issues of poverty, and the poor are most vulnerable to the impact of climate change, which evidently requires adaptation policies and bottom-up approaches.

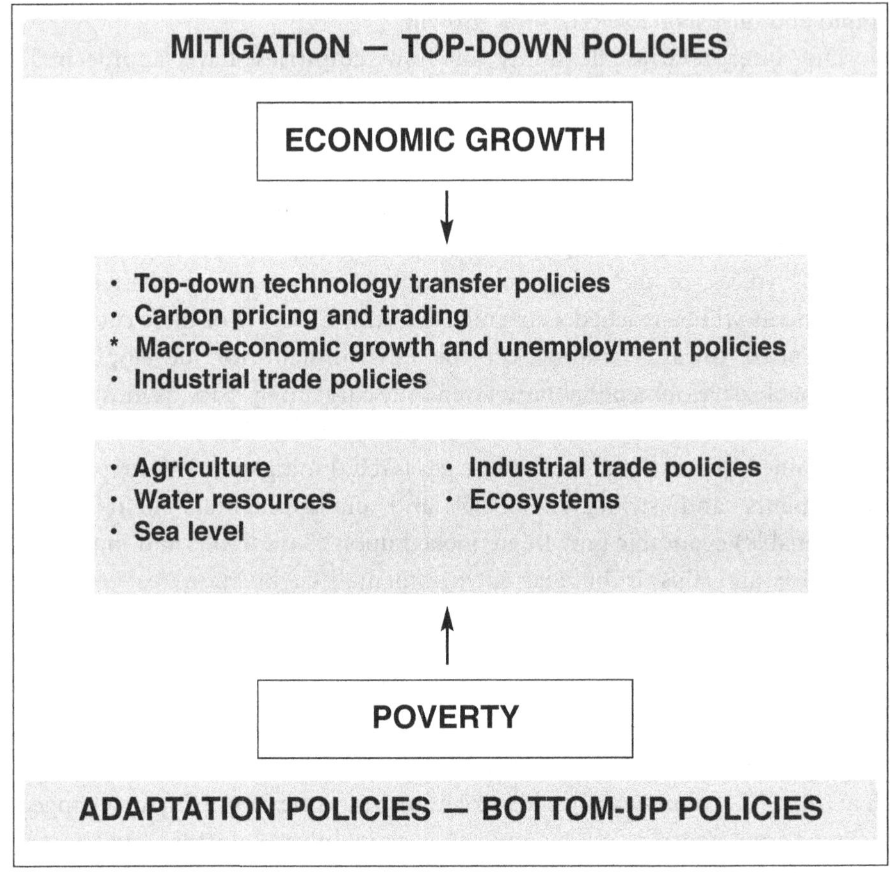

Source: Author's Own

ADAPTATION – THE MISSING DISCOURSE

Adaptation[19] as a climate change policy was only put on the international agenda in the 2000s. However, mitigation and climate financing continue to dominate the climate change agenda (Death, 2014; Madzwamuse, 2014) and adaptation has been ignored out of fear of admitting defeat and

19. The IPCC defines adaptation as the adjustment in natural or human systems to a new or changing environment, and also involves efforts to limit our vulnerability to climate change. It defines mitigation as an anthropogenic intervention to reduce the sources or enhance the sinks of greenhouse gases (GHG).

undermining support for the Kyoto framework (Roos, 2010). According to Christoplos (2014), the green growth[20] rhetoric suggests the existence of 'win-win' solutions whilst, in truth, adaptation to climate change has been a more important but under-discussed challenge for many developing countries. Political and climate policy advocates from developing countries have shied away from discussing adaptation or the issues of people who are most vulnerable to climate change due to a belief that to do so would weaken political resolve to pursue effective mitigation (Thomson and Rayner, 1998).

Figure 1 shows the disparate policy inclinations of the mitigation and adaptation agendas of the United Nations Framework Convention on Climate Change (UNFCCC), showing the heavy bias of mitigation policies towards pro-growth agendas. Little attention has been paid to adaptation in international and national climate change discourses and politics even though this is where a large part of the focus of poor countries in Africa and the BASIC countries should be.

Should the scale and the absence of modes of production in African countries, similar to traditional industrialisation, which make them less pollutive than other emerging and developed countries, be seen as an advantage? Africa has more projects to protect biodiversity than other major areas of the world and these projects are vulnerable to the effects of climate change. Countries in Africa are particularly vulnerable to climate change and tend to have a much higher share of their economy in climate-sensitive sectors (Smith and Lenhart, 1996).[21]

Some of the reasons for the heavy focus on mitigation can be enumerated as follows: policies on adaptation are looked upon as substitutes for mitigation; the climate mitigation strategies instituted at the UNFCCC serve market-based economies and legitimise economic and technological systems of the countries of the North; and adaptation requires local and improved and interactive institutional and political capacities to deal with climate change, which the UNFCCC framework had, for long, largely ignored.

20. Death (2014) discusses four types of Green Economy discourses: Green Revolution; Green Transformation; Green Growth; and Green Resilience. Although evidence of the four discourses can be found in South Africa, different state actors and institutions have different positions on the green economy; the green growth discourse is the most embedded and most widespread of these discourses, articulated as such in the 2011 White Paper on Climate Change, the Green Economy Accord, and National Development Plan (NDP), among others.

21. See Smith and Lenhart (1996) for a description of a number of government policies that could be implemented in anticipation of climate change to reduce its potential adverse effects. The policy options are divided into five climate-sensitive sectors: water resources, coastal resources (sea-level rise), forests, ecosystems, and agriculture.

THE ETHICS OF MITIGATING CONDITIONS: DO WE NEED GREENING TECHNOLOGIES?

One of the main pillars of the UNFCCC has been the importance given to technology transfer. The 2007 Bali Action Plan identified technology transfer areas as one of the four priority areas and Article 4.5 of the Convention explicitly states that, for developing countries to successfully implement the provisions of the UNFCCC, developed countries would be required to transfer environmentally sound technologies and know-how to developing countries (Latif, 2014). The Stern Review also emphasises the need for, and importance of, improved technology and, as such, technological progress can help reduce GHG emissions (Stern, 2007).

It is quite legitimate for developing countries to make the implementation of the provisions contingent upon the transfer of environmentally sound technologies (ESTs) from developed countries, and to argue that there is a moral and ethical imperative for them to be given access to technologies to help combat climate change, but this has led to ambiguous sustainability goals. The question has been posed, though, as to whether this can be pursued without:

The Undermining of Responsibility by Developing Countries to Take any Necessary Action

At the 2009 COP in Copenhagen, South Africa announced a 34 per cent reduction in emissions by 2020 below business-as-usual, with emissions expected to level out and decline in real terms from 2036. But the government added that the achievement of these national targets is subject to the availability of significant financial, technological and other support from the international community or developed countries. Other emerging countries, such as India, have stated that a lack of access to improved technologies and ESTs is going to become a major impediment in combating climate change.

Yet, developing countries have not detailed or explained what kind of technologies would help combat climate change and how. There has to be sufficient knowledge and absorptive capacity in the technology recipient countries to be able to adequately use and adapt the ESTs. Moreover, no technology can be indiscriminately applied to every socio-economic context, and it is also critical to understand how the use of a new technology will impact the distribution and use of land, water and other resources (Christoplos, 2014). Emphasis has been placed on the transfer of those technologies that continue to sustain economic growth and those that have large-scale and commercial applications such as carbon capture and storage (CCS), shale gas fracking and large-scale wind and solar projects. While these technologies do have some positive impact, without people-centred approaches they are often of little advantage to people most affected by climate change – the poor who constitute the majority of the population in countries such as India, China, South Africa, Brazil and other African countries.

The Stalemating and Creation of Deadlock Situations in Climate Negotiations

Further, developing countries have to clearly flesh out how intellectual property rights (IPRs) for environmentally sound and clean energy technologies are a barrier to combating climate change (Latif, 2014). Many companies from developed countries are strongly opposed to sharing their IPRs. Yet, in such discussions, there needs to be an appreciation of the role of IPRs as essential incentives to promote innovation. Perhaps, as with the negotiations, cajoling and even defiance that characterised the debate around pharmaceuticals over the past few decades, a new equilibrium needs to be established in this relationship.

Thus, a middle ground in the UNFCCC discussions around technology transfer may be found and current deadlocks broken. Latif (2014) provides a number of parameters to help structure the discussion on this complex issue and thus avoid a deadlock: minimise the importance of IPRs while giving more importance to an enabling environment, finance and absorptive capacity; scrutinise litigation cases between companies of the North and the South to understand the implications of the technology transfer of climate change technologies, and whether IPR is actually an issue as it has not prevented the emergence of industry champions in China and India; and draw on practices and lessons of existing bilateral R&D and technology partnerships between developed and developing countries.

RECOMMENDATIONS AND CONCLUSIONS

Political arguments that a green economy and green growth will create new jobs and overcome poverty are often overstated in relation to a real transition towards sustainable development. Such statements unfortunately only legitimise market-led principles of economic growth that continue to support modes of production and consumption that are often not energy-saving, whilst taking all attention away from those elements that are in serious need of contemplation and critical policy intervention, such as grassroots and informal sectors of the economy. For developing countries such as the BASIC group, there should be a serious focus on linking sustainable development and climate change to poverty.

The world is in need of a climate change framework for transnational and global political action that embraces new and alternative thoughts and action. Examples of these are those proposed by de-growth and the radical ecological democracy proponents, which put collectives and communities at the centre of sustainable governance and the economy, and provide a holistic vision of human well-being encompassing physical, material and sociocultural dimensions. Such thinking requires the spread of certain core values of sustainable development, and the sharing of these values among countries.

It is these alternative frameworks and modes of thinking and action that will unlock technological, institutional and economic systems from old habits and thinking. Now is the time to scrutinise the specific ways in which the current world order has been created, the kind of thinking and practices that have led to its failings, and the systemic interventions required to extricate humanity from the current malaise.

There is a need for framing the issue of sustainable development in a manner that embraces the social complexity or the human and multidimensional nature of the issues. That requires a revolutionary change in political consciousness within and across national boundaries, including a keen appreciation of the conditions of the poor and marginalised across the globe. Domestic and national climate change and sustainable development policies of at least the major countries in the North and the South must be embedded within a broader transparent international effort and discourse.

Both at the UNFCCC and at the national levels of the BASIC countries, the important issue of adaptation as a climate change policy has been an under-discussed challenge. It has, however, gained ground since COP 20 in December 2014, and will be one of the critical proposals in the agenda for COP21 in December 2015. This must be sustained and intensified, taking into account the fact that the poor, especially in Africa, Asia and Latin America, are in the majority and extensively dependent on climate-sensitive sectors for their sustenance.

References

Ahmed, F. 2004. 'An Examination of the Development Path Taken by Small Island Developing States: Jamaica a Case Study', Master's thesis. Canada: the University of Western Ontario.

Athanasiou, T. 2014. 'Climate Crossroads: Toward a Just Deal in Paris', *A Great Transition Initiative* Viewpoint. Available at: www.greattransition.org [Accessed: 20 March 2015].

Christoplos, I. 2014. 'The Downside of Green Growth, Technological Choices'. Available at www.danc.eu [Accessed: 28 March 2015].

Davis, R. 2015. 'A Meaningful Step on a Long and Winding Road' in Ed Wallis (ed.) *Making a Global Deal on Climate Change a Reality*. Fabian Society. Chapter 1, pp. 1–8.

Death, C. 2014. 'The Green Economy in South Africa: Global Discourses and Local Politics', *Politikon: South African Journal of Political Studies*, 41(1) 1–22.

Falk, R. 2014. 'Changing the Political Climate: A Transitional Imperative', *A Great Transition Initiative* Essay. Available at: www.greattransition.org [Accessed: 16 March 2015].

IAC. 2010. 'Evaluation of IPCC's assessment process, climate change assessments: a review of the processes and procedures of the IPCC'. Available at www.interacademycouncil.net [Accessed: 30 March 2015].

Kallis, G. 2015. 'The Degrowth Alternative', *A Great Transition Initiative* Viewpoint. Available at: www.greattransition.org [Accessed: 20 March 2015].

Kothari, A. 2014. 'Radical Ecological Democracy: A Path Forward for India and Beyond', *A Great Transition Initiative* Essay. Available at: www.greattransition.org [Accessed: 20 March 2015].

Lippert, I. 2004. 'An Introduction to the Criticism on Sustainable Development'. Available at: www.academia.edu [Accessed: 10 May 2015].

Lütkenhorst, W., Altenburg, T., Pegels, A. and Vidican, G. 2014. 'Green Industrial Policy: Managing Transformation Under Uncertainty'. Discussion Paper 28/2014. German Development Institute.

Madzwamuse, M. 2014. 'Drowning Voices: The Climate Change Discourse in South Africa'. Heinrich Boll Stiftung-Southern Africa, 3 February. Available at: http://za.boell.org/2014/02/03/drowning-voices-climate-change-discourse-south-africa-publications [Accessed: 10 April 2015].

Mahapatra, B. 2014. 'The Modi Government's War on Environment', 12 November 2014, *Open India*. Available at: www.opendemocracy.net [Accessed: 10 April 2015].

McCarthy, S. 2015. 'UN panel calls Canada a climate laggard'. Available at: www.theglobeandmail.com [Accessed: 20 April 2015].

Our Common Future, Chapter 3: 'The Role of the International Economy', Article 27, taken from A/42/427, *Our Common Future: Report of the World Commission on Environment and Development*. Available at: www.un-documents.net [Accessed: 8 August 2015].

Pearce, F. 2010. *The Coming Population Crash: and Our Planet's Surprising Future*. Boston, Massachusetts: Beacon.

Pearce, F. 2014. 'It's Not Overpopulation That Causes Climate Change, It's Overconsumption', *The Guardian*, 9 September 2014. Available at: www.theguardian.com. [Accessed: 24 May 2015].

Perrot, R. and Summoni, D. 2015. 'A Critique of the Green Economy in Africa', *Foresight for Development*. Available at: www.foresightfordevelopment.org [Accessed: 20 April 2015].

Roos, J. 2010. 'How to Eat an Elephant: A Bottom-Up Approach to Climate Policy', The Breakthrough Institute. Available at: www.thebreakthrough.org [Accessed: 10 May 2015].

Smith, J. B. and Lenhart, S. S. 1996. 'Climate Change Adaptation Policy Options', *Climate Research*, 6(2), 193–201.

Stollery, K. R. 1983. 'Mineral Depletion with Cost as the Extraction Limit: A Model Applied to the Behaviour of Prices in the Nickel Industry', *Journal of Environmental Economics and Management*, 10: 151–165.

The Hague Report. 1991. 'Sustainable Development: From Concept to Action', Dutch Ministry of Development Cooperation.

Voß, J. P. and Kemp, R. 2006. 'Sustainability and Reflexive Governance: Introduction', *Reflexive Governance for Sustainable Development*, 3–28.

CHAPTER 2

LTMS and Environmental and Energy Policy Planning in South Africa: Betwixt Utopia and Dystopia

Louise Scholtz, Manisha Gulati and Saliem Fakir

Only a crisis – actual or perceived – produces real change. When that crisis occurs, the actions that are taken depend on the ideas that are lying around. That, I believe, is our basic function: to develop alternatives to existing policies, to keep them alive and available until the politically impossible becomes the politically inevitable (Milton Friedman, *Capitalism and Freedom*, 1962, cited in Seguiono, 2014: 8).

It is ... impossible to draw a clear line between energy resources, climate policies and geopolitics (De Micco, *et al.* 2013).

INTRODUCTION

While there is consensus that the environmental challenges facing the world will require a major transformation, 'widely differing interpretations exist about the nature of this transformation, with competing assumptions about the institutions required and the drivers of change (Cook, *et al.* 2012: iii).'

Any transition to a low carbon trajectory is highly dependent, not only on policies and legislation, but also on the level of acceptance and support by the status quo for the envisaged transition. As such, both existing and 'future power relations, governance arrangements, and participatory processes' are central to envisioning and implementing the transition (Cook, *et al.* 2012). Related to this is the debate on whether existing 'democratic forms of governance' will be able to drive the urgent changes in policy required or whether more coercive and autocratic governments may be necessary (Held and Hervey, 2012: 5). In addition, Scholvin (2014: 186) alerts that, though economic and political causes are important, one should not lose sight of how natural conditions and endowments direct policy. 'Nature is linked with path dependency', demonstrated by the extent to which historical energy policies translated into 'material structures in geographical space limit the policy options available today.'

In the case of South Africa, an ample supply of coal is both the main determinant of existing energy policies and the obstacle to a transition to a low-carbon economy. South Africa is a carbon-intensive economy that derives 86 per cent of its electricity from coal (Fakir and Gulati, 2012), emits 520–540 million megatons of carbon a year (estimates for 2010 based on the inventory for the year 2000) (Marquard, Trollip and Winkler, 2011), contributes 1.49 per cent of total global CO_2 emissions, and with per capita emissions of nine tons in 2005, compared to an average for Annex I countries of 14 tons, for non-Annex-I countries[22] of 4.8 tons, and a world average of 6.7 tons (Marquard, Trollip and Winkler, 2011). Presently, the country is in twelfth place on the list of carbon emitters in the world and first in Africa (Satgar, 2014). Compounding matters is the country's minerals-energy complex (MEC) that is still central to the South African economy, raising questions as to whether climate mitigation can be seen as a coherent project in South Africa.

Nevertheless, the government has taken steps towards an effective climate change response with regards to emissions reductions and the long-term transition to a climate-resilient, equitable and internationally competitive low-carbon economy. These steps first found expression in the Long-term Mitigation Scenarios (LTMS). Adopted in 2008, the LTMS explores the

22. Annex-I countries refers to industrialised countries that were members of the Organisation for Economic Co-operation and Development (OECD) in 1992, plus countries with economies in transition including the Russian Federation, the Baltic States, and several Central and Eastern European States that have ratified or acceded to the United Nations Convention on Climate Change, which sets an overall framework for intergovernmental efforts to tackle the challenge posed by climate change; Non-Annex-I countries are mostly developing countries that have ratified or acceded to the United Nations Convention on Climate Change but are not included in Annex I (United Nations Framework Convention on Climate Change, undated).

country's options for climate change mitigation. It overlays a number of possible emissions trajectories, from the (unrestricted) Business as Usual (BAU) at the top, through reductions expected from three sets of interventions improving on current development plans, to bringing the country broadly in line with reductions required within an effective international climate change mitigation effort. The central question this chapter seeks to address is whether the LTMS and related policies adequately address climate change and support and drive the transition to a low-carbon economy in South Africa. The chapter (in the main) argues that the LTMS, in particular, and energy and environmental policy, as a whole, was developed in a manner that largely ignores 'data from the real world' such as vested interests, the political economy and the critical role of coal as key driver of the economy (Sequino, 2014: 1), and is hampered by policy dissonance.

By highlighting the challenges of environmental and energy planning in the context of the LTMS, this chapter aims to provide a framing for some of the debates in the book. A wider analysis of the climate policies of seven countries provides a conceptual framework to understanding the levers that inform 'coherent' climate policy.

The first part of this chapter delves down into the relevant South African policies and the LTMS in particular, i.e. it's role in policymaking, and its limitations. The second part sets out the challenges to environmental and energy planning in view of these shortcomings. Part 3 highlights approaches by specific countries and commonalities relevant to South Africa, while the conclusion alerts to opportunities and challenges that can impact on South Africa's low-carbon future.

PART 1: SOUTH AFRICA'S LONG-TERM MITIGATION SCENARIOS

Overview of LTMS

South Africa's comprehensive response to climate change mitigation interventions began with the LTMS document, which assesses the mitigation potential and opportunities for the country and purports to be the basis for an informed policy on tackling climate change. The LTMS lays down a number of emissions reduction scenarios and trajectories for the period 2003 to 2050 (Figure 1). These scenarios include a 'Growth Without Constraints' (GWC) scenario, 'Required By Science' (RBS) scenario, and

Figure 1: LTMS modelled and proposed emissions trajectories

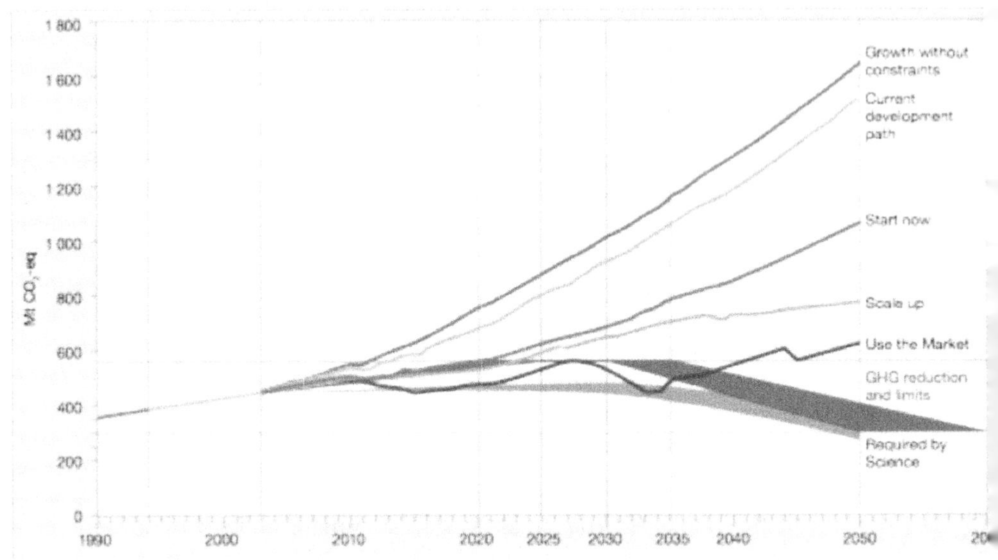

Source: DEAT 2007 in WWF 2011

several intermediate scenarios or strategic options for mitigation that the country can embark upon.

Emissions projections presented in the LTMS report – approved by the South African cabinet in July 2008 – display indicative Peak, Plateau and Decline projections (PPD) (shown in red). In brown are the LTMS projections of 'Growth without Constraints' and a 'Required by Science' pathway that assumes considerable latitude for developing countries; and three scenarios based on quantified packages of mitigation actions: 1) Start now, 2) Scale up and 3) Use the market.

The GWC scenario is simply a scenario in which things continue as they are: i.e. BAU. This scenario outlines the country's emissions if energy plans continue to be based purely on least-cost, without any constraints on growth or carbon. It is based on the assumption that the country's energy economy continues to evolve around the minerals-energy complex (MEC) and there are no climate impacts highly damaging to the economy. By the very nature of assumptions and absence of accounting for externalities, this scenario is based on fossil fuel options that were cheaper at the time the LTMS processes were undertaken and not implementing a carbon budget. It suggests a quadrupling of carbon emissions from 440 Mt CO_2-eq in 2003 to around 1600 Mt CO_2-eq in 2050.

On the other hand, the RBS scenario was guided by the recommendations of the Intergovernmental Panel on Climate Change (IPCC). This is a notional scenario that assumes that the country undertakes mitigation measures to the extent required by science to meet its fair contribution towards global emission reductions. The underlying premise is that absolute carbon outputs from 1990 levels should be reduced by 60–80 per cent by 2050 through embarking on low-carbon technology trajectories. This scenario suggests a peaking of emissions around 2020, and slow decline to between 240 Mt CO_2-eq and 260 Mt CO_2-eq.

A comparison of the GWC and RBS scenarios suggests a difference of around 1300 Mt CO_2-eq in 2050. This gap is more than three times the annual emissions in 2003. This difference in emissions would be the mitigation cap that would be imposed on the country in the event of strict adherence to the RBS scenario. In other words, adherence to the RBS necessitates cost-effective low-carbon solutions that do not allow the emissions of an additional 1300 Mt CO_2-eq per annum if the country needs to grow without constraints.

The LTMS provides the underpinning for informed debate about South Africa's future development. In doing so it surfaces three important issues. First, it conveys the recognition that high carbon and energy intensity is not a sustainable and flexible pathway for South Africa's economy in the long-term because increased dependence on carbon-intense fossil fuels will show diminishing benefits with time. Second, it shows that significant shifts can be made in energy and transport sectors. (Issues of affordability and pace of scale will be key to the nature of change and transition.) Third, low-carbon technology trajectories and markets offer economic opportunities that can reduce system risks to external shocks, inflation and price volatility. Simultaneously, they have the potential to create and generate new types of economic sectors and growth levers.

LTMS and Policymaking

Relying on the LTMS, South Africa's policy framework has committed to climate change mitigation and adaptation interventions in the country being informed by, and monitored and measured against, a PPD emission trajectory where GHG emissions peak in 2020–25, plateau for approximately a decade, and decline in absolute terms thereafter. The RBS scenario also informed the country's bold and ambitious targets announced in Copenhagen in 2009. These targets (referred to as the Copenhagen Commitment) involve a reduction in the country's carbon emissions by 34 per cent in 2020, and 42 per cent by 2025, conditional on finance, technology and capacity-building support from the international community.

The LTMS findings have given important impetus to the 2011 National Climate Change Response White Paper, which is the first coherent outline of national government's responsibilities relating to mitigation and adaptation and is the basis for future government action. The strategy laid down in this paper adopts a strongly sectoral approach, emphasising immediate implementation of 'Near-term Priority Flagship Programmes' focusing both on reducing emissions and on adaptation (in an approach it characterises as 'climate change resilient development').

Subsequently, a number of steps have been taken to incorporate climate change into national and sub-national policies. A carbon constraint has been introduced into energy planning; a large-scale renewables power purchase programme has been initiated; policy measures have been introduced to incentivise the uptake of energy efficiency; levies such as a vehicle carbon emissions tax and a tax on incandescent light bulbs have been introduced; and a carbon tax is proposed to be levied from 2016 (National Treasury, Republic of South Africa, 2014).

The government has undertaken a mitigation potential analysis (MPA) for key sectors to analyse emissions mitigation options. The MPA is being used to identify the short- (2016–2020), medium- (2030) and long-term (2050) desired emission reduction outcomes (DEROs) for each significant sector (or, where appropriate, subsector) of the economy, and where necessary translate these into carbon budgets for companies or entities in the said sector. The determination of DEROs involves the development of an appropriate mix of policy measures per sector (Mix of Measures), or at subsector level, to enable, support, catalyse or regulate the DEROs (See Table 1 for a policy overview).

The government has also embarked on the process of developing long-term adaptation scenarios to inform adaptation planning and implementation. This process is expected to project and evaluate the socio-enviro-economic implications of the potential impacts of climate change and climate variability for sectors such as water, agriculture, biodiversity, health, forestry and fisheries, and human settlements.

Table 1: Broad overview of South Africa's Climate Policy Framework

Agency/Department	Policy	Thrust
The Presidency: National Planning Commission	National Development Plan (NDP) 2030	Includes climate change as part of the long-term development planning process, market and private sector-led mitigation and adaptation approach to climate change.
Economic Development	National Growth Path	Climate change as part of green economy, green jobs and broader development strategy.
National Treasury	Market-based instruments	Carbon tax, carbon offsets, environmental levies, energy efficiency tax credits and procurement of renewable energy.
Environmental Affairs	Long-term Mitigation Scenarios; National Climate Change Response Paper; Green Fund	Peak, plateau and decline approach to carbon emissions. Some level of government coordination together with market instruments for mitigation and adaptation. Financing green projects and research.
Energy	White Paper on Renewable Energy and Renewable Energy Independent Power Producer Procurement (REIPPP); Integrated Resource Plan for Electricity 2010–2030; Integrated Energy Plan; Renewable Energy Independent Power Producer Procurement Programme	Energy mix and consumption as responsible for current GHG emissions.
Cooperative Governance and Traditional Affairs		Oversight of implementation of climate-related measures by provincial and local governments and legislative power on climate-related issues. Responsible for disaster management.
Trade and Industry	Green Economy Accord; Industrial Policy Action Plan	Greening of industrial policy; solar water heater roll-out

Source: Compiled by authors from Satgar, 2014; National Treasury 2013; Giordano, *et al.* 2011

Why is the LTMS not Working?

The LTMS provides a good overview of the scale of the climate change mitigation challenge in South Africa; the options (technology, policy and market instruments) to address it; and the economic costs and benefits. In short, it furnishes the base for a robust long-term national climate policy. Regrettably, neither the LTMS nor the policies it inspired seem to be working. There are several reasons for this.

First, the LTMS is focused on identifying least-cost mitigation options rather than on developing economically and politically viable climate compatible development options, leaving a gap between its findings and the development of implementable policies. For example, the electricity supply crisis of 2007–08 (arising from a lack of planning and investment in the sector), which abruptly ended decades of low-cost and abundant electricity supply, and the resulting unreliable provision of affordable electricity, has been cited as one of the biggest constraints to economic and social development in the country. Consequently, security of energy supply and cheap electricity has dominated the policy agenda, often at the expense of the environment.

Second, the proposed scenarios have not sufficiently tested the ground on developmental impacts embedded in trade-offs, the ability of the South African economy or specific sectors to lock out of current fuel and technology pathways, and terms of trade. This is pertinent, given the country's MEC that lies at the 'core of the South African economy, not only by virtue of its weight in economic activity but also through its determining role throughout the rest of the economy' (Fine and Rustomjee, 1996).

Although the economic model based around the MEC, which previously provided domestic and foreign capital with cheap and plentiful coal-generated electricity, is deemed no longer economically or environmentally sustainable (Baker, Newell and Phillips, 2014), the MEC continues to be at the heart of the South African economy with its interlinkages in finance, manufacturing, service industries and other sectors. The MEC collectively accounts for close to 20 per cent of South Africa's GDP (NPC 2013 and Fine and Rustomjee, 1996). It is the largest source of export earnings and is responsible for 500,000 direct jobs. It is expected to continue to play an important role in the economy, with some estimates (NPC 2013) suggesting that its contribution to GDP can double by 2030 and it could add another 200,000 jobs.

It is argued that the lack of good data sets and tools renders it difficult to fully consider development impacts. The available data sets, though the best available, were inadequate because the sort of data required for such an exercise did not exist in the country. After all, this was the first time that the country was doing such an exercise. Nevertheless, it is important to recognise that the above-mentioned economic realities constrain the country's abilities to embark on a low-carbon transition on the basis of an RBS scenario even though this goal has been somewhat modified as a result of South Africa's international commitments.

Third, the LTMS is naïve about the 'political settlement',[23] i.e. the evolution, distribution and structure of social power, which profoundly influences and impacts on the likelihood of a particular policy achieving its expected outcomes, and it does not support the low-carbon transition. In South Africa's case, the political settlement is dominated by the carbon-intensive MEC. The constellation of class forces underpinning the MEC, spanning and joining elites (party-state-business), regularly and repeatedly employs networks and connections to influence and shape praxis in ways that protect the core MEC sectors (see Baker, 2013). This network works tirelessly to modify the organisational conditions governing the generation and allocation of rents.[24]

Climate-related policy intervention changes income flows of environmentally damaging companies negatively, and creates negative rents for such companies. Thus, by their very nature, these policies create conditions where the carbon-intensive MEC in South Africa loses rents. Additionally, it changes the conditions that the MEC has to fulfil to maintain or gain rents. This is because the MEC has to invest in new technology or more efficient modes of production to lower its environmental impact.

Some other aspects of the political settlement are the mode of policymaking and the structure of the electricity sector in the country. In terms of climate-related policymaking, there is no centralised lead agency to directly control environmental matters in an integrated fashion. At least 21 of the 32 sectoral departments are concerned with climate change. The LTMS does not spell out the actual possible combination of approaches, policies, measures and actions to achieve optimal mitigation or how to dovetail

23. The political settlement refers to the configuration of capabilities and distribution of organisational and bargaining power across economic, political and bureaucratic organisations in a society (Khan, 2010). Thus, a political settlement includes governance agencies or the bureaucracies within the state that are charged with the management of policies to monitor and enforce policies. The technical capabilities of governance agencies and the relative power of these organisations relative to other organisations jointly determine the ability of governance agencies to enforce policies (Khan, 2013).

24. Rents are defined as incomes higher than the minimum a person or organisation would have accepted, the minimum usually being defined as the income in the next-best opportunity available to that individual or organisation (Khan, 2013).

different policy measures such as taxes, emissions reduction targets, efficiency targets, standards and regulatory measures. This is left to the implementing government departments and agencies. The DEROs process could achieve such coordination. However, experience with this process and discussions with experts and stakeholders involved suggests otherwise. In the absence of combinations and dovetailing, the resulting ineffectiveness and poor targeting is of little or no surprise.

In the electricity sector there are structural supply problems. The sector is characterised by the absence of competition and supply diversity at the wholesale or retail level. End-users have no choice in terms of energy supplier or technology and cannot switch to another supplier with a lower carbon footprint.

Fourth, though the LTMS is a useful introduction to technology options for mitigation, these options do not adequately address the cost, scale and institutional barriers that (either) limit options or the pace at which a particular technology pathway can be embarked upon. Technologies, institutions, business strategies and user practices have evolved in a manner that has locked the country into an energy and carbon-intensive economic pathway. This lock-in is systemic and mutually reinforcing: the technology lock-in delays the adoption of more efficient technologies thereby causing the institutional lock-in and the institutional lock-in makes it difficult to direct investment to new, more carbon-efficient classes of infrastructure.

Fifth, the LTMS lacks a techno-economic assessment and feasibility stress test in terms of the ability of different sectors to adjust to emissions reduction targets. Take the example of cement, which is central to the National Development Plan (NDP), as well as the National Infrastructure Development Plan, the New Growth Path and the Industrial Policy and Action Plan. The cement production process is inherently resource intensive. Cement is a low-value, high-volume material with a high material and energy throughput. Clinker production (i.e. calcination of limestone) accounts for around 50 per cent of the total emissions from cement manufacturing. The percentage of clinker substitutes as a percentage of the total mix of materials in South Africa has increased from 12 per cent in 1990 through 23 per cent in 2000 to 41 per cent in 2009. Technology and the burning process, which remained constant for a number of years, prevent further emissions reduction possibilities. New investments in terms of technological upgrades to improve efficiency and reduce emissions can range between R50–60 million. Given the age of the plants and the falling demand for cement in the country, such investment is not justified. It would be better, some argue, to close down these plants (Scholtz and Gulati, unpublished).

Future emissions reduction opportunities for the South African cement industry range from waste heat recovery (WHR) to the use of waste and secondary materials in kilns, carbon capture and storage (CCS), and use of innovative cement materials. However, each of these poses challenges. Novel cements using innovative materials are limited both by limited proven long-term durability and economic viability. Given that cement is a homogenous product and that there is very little product differentiation, and products from different producers can generally be substituted for each other, there is no advantage in producing lower-emissions-intensive products at higher costs. CCS is limited by substantially high capital and operational costs. The exact capital expenditure needed per plant is difficult to determine at present, but is estimated by the European Cement Research Academy to be around €330–360 million to deploy oxyfuel technology at a new one million tonne/year plant, about €100 million for retrofitting oxyfuel technology and €100–300 million to retrofit an existing plant with post-combustion technology (Scholtz and Gulati, unpublished). Operational costs of a plant equipped with post-combustion carbon capture technology are estimated to be double the cost of a conventional cement plant, while oxyfuel use would incur 25 per cent higher operating costs (Scholtz and Gulati, unpublished).

Similarly, WHR may be limited by the initial investment costs and a long payback period. The use of waste and secondary materials in kilns is cost effective and the resulting increase in energy efficiency enhances the business case for WHR with the continuing rise in electricity prices in the country. But problems such as the absence of assured and affordable supply chains of waste, lack of waste segregation, and poor economics make this option more expensive and cumbersome than burning coal.

Sixth, the LTMS is still to grapple with whether South Africa is best as a technology leader or as an adopter. Evidence seems to suggest that its capabilities are best placed in the spectrum of rapid adoption rather than being low-carbon technology leaders. This is of particular significance because a technology leader requires access to a highly skilled workforce, state wherewithal for long Research and Development (R&D) horizons, incentives, and ability to mobilise (risk averse) pre-market finance that can carry new technologies and innovations beyond the 'Valley of Death'.[25]

25. 'Valley of Death' is a term used to describe a stage in technology development that lies between the stage when a technology has been financed at the laboratory, or even pilot stage, and when it is rolled out commercially on a large scale. In this stage, a technology is too capital intensive for a venture capital firm to continue investing, but too risky for a bank or private equity firm to bring it to scale.

Given South Africa's historical 'entanglement with mining and fossil fuels' (Rafey, 2013) and technological lock-in into fossil fuel based and carbon-intensive technology and infrastructure at the systemic level, innovation has tended to focus on using fossil fuels (more) efficiently. The South African energy innovation system is locked into fossil fuels and has resulted in the country becoming a world leader in fossil-fuel-based technologies (Scholtz and Gulati, 2013). One such example is Sasol's coal-to-liquid fuel plants, a technology commercialised nowhere else in the world. The NDP 2030 reinforces this – Satgar (2014: 143), for example, points out that many of the 'techno fixes' referred to in the Plan focus on the reduction of the carbon footprint by existing and future coal-powered power stations through retrofitting, clean coal technologies and further investigation of the feasibility of carbon capture and storage. Against this backdrop, South Africa has not innovated on low-carbon technologies and has been adopting technologies developed elsewhere.

At the same time, however, 'borrowing and applying frontier technologies' is often difficult, and implies export capability to earn foreign exchange to do so. Government subsidies – via the provision of 'economic rents' – can potentially deliver cost containment. By keeping costs low, companies can be incentivised to migrate to new industries in which they have little experience. Additional tools could include cheap credit and the promise of bailout support (Sequino, 2014: 4, 5). All of these would present a challenge and prove problematic to the MEC dominated political settlement.

Finally, the LTMS does not address the issue of the country's absorptive capacities in respect of climate friendly technologies. These capacities pertain to skills required to operate and maintain new technologies but also include the motivating or persuasion factors for end-users to adopt these technologies.

PART 2: CHALLENGES TO ENVIRONMENTAL AND ENERGY PLANNING IN VIEW OF THESE SHORTCOMINGS

The above-identified limitations of the LTMS make pursuing environmentally friendly energy planning significantly more challenging. Additionally, the multitude of implementing agencies and diverse sectoral plans that target low carbon development with little integration amongst them has meant that different plans have different visions and approaches to achieve the objectives of economic development and low-carbon transition. For example, while the Industrial Policy Action Plan and the 10-year innovation plan envisage a skills and innovation-based economic transition, the energy plan adopts a traditional infrastructural-driven approach to the transition and targets economic development through exploitation of natural resources, especially coal and its export. Similarly, there has been much debate over the link between the tax and the carbon budget. A carbon budget is not fundamentally different from a cap on emissions. A process of carbon budgeting involves the allocation of carbon, which is a limited resource. The viability of placing a cap on emissions, while at the same time taxing emissions, is therefore debatable. There is, however, no clarity on how the carbon tax and the carbon budget are going to interact.

Then there is the issue of sufficient will or capacity on the part of the government to radically transform the South African economy in a more environmentally sustainable direction (Death, 2014: 19). A similar question relates to the extent to which it is responsive to the social, economic and political realities of South Africa, its natural endowments and development needs. These disconnects can be illustrated through examples from policymaking across sectors in the country.

The first example is that of electricity planning. Given that coal accounts for 86 per cent of the country's electricity generation – thereby making the electricity sector account for half of the country's emissions – the electricity system is a key point of intervention from the climate change and lower-carbon perspective. Yet, planning for electricity generation capacity expansion, as conducted in the form of the Integrated Resource Plan (IRP), continues to be overwhelmingly anchored in least-cost optimisation (Burton and Coetzee, undated). Consequently, in absolute terms, coal increases by 6,000 MW to a total of 50,771 MW (even though proportionally it decreases from 85 per cent to 65 per cent). Moreover, demand forecasts presuppose the ongoing dependence of the economy on energy-intensive industry (Burton and Coetzee, undated).

It is estimated that 11,000 of the 16,000 MW of the new coal-fired plant in the IRP goes towards meeting forecast energy demand from industry (Trollip and Tyler 2011 in Burton and Coetzee). In fact, the final IRP states that the policy requirement for continuing a coal programme could result in this coal-fired generation (new coal to be built by 2026) being brought forward.

Existing coal-fired generation is run at lower load factors[26] to accommodate the new coal options while the emissions target applies. This suggests an inefficient use of the electricity system, so as to meet emissions constraints imposed on the model as well as meet the goals of a 'coal policy' that has never been clearly delineated by government (Burton and Coetzee, undated). (Again, this is a clear case of the 'political settlement' influencing electricity planning and the mix of technologies therein.) The other side of the coin in this case could be to say that the additional coal-fired power proposed in the IRP is the plateau in the PPD emissions trajectory. However, the IRP has not provided for such a technical analysis.

The update to the IRP 2030 that is currently under discussion reinforces the fixation with coal. It suggests that regional coal options are attractive due to the emissions not accruing to South Africa, and in cases where the pricing is competitive with South African options, regional coal would be preferred (Department of Energy, Republic of South Africa, 2013). It then suggests that alternatives to extending the life of the existing fleet of coal-fired power plants would be to build new coal-fired generation that is more efficient and has lower emission rates, and non-emitting alternatives would be considered under more aggressive climate mitigation objectives (Department of Energy, Republic of South Africa, 2013). However, there is no discussion of what more aggressive climate mitigation objectives would mean. Similarly, the least-cost options are emphasised throughout the proposed revisions. The update states that: if the price of power from hydro power plants, inclusive of the transmission requirements to evacuate the power from the plants, can be maintained within the 60 cents per kWh benchmark for new coal, then these options should be pursued as and when they arise (Department of Energy, Republic of South Africa, 2013).

Finally, the update acknowledges the shortage of electricity supply until 2016 and states that the only options available to manage the situation are increased energy efficiency and demand-side responses, and improved utilisation of existing generation resources (Department of Energy, Republic of South Africa, 2013). The latter refers to the improvement of Eskom's fleet performance and incentivising production from existing non-Eskom generation. It completely ignores the role that distributed power solutions using renewables can play in alleviating the shortage and meeting the electricity needs of the people.

26. The load factor of a power plant is the ratio of a power plant's actual output over a period of time to the maximum possible output that can be produced by the power plant if it operated to its full capacity continuously over the same period of time.

The second example comes from the Strategic Integrated Projects (SIP). The NDP for South Africa incorporates the National Infrastructure Development Plan (mentioned above). The Infrastructure Plan lists a series of ambitious and far-reaching initiatives aimed at transforming South Africa's economic landscape, addressing unemployment and improving basic service delivery. To this end, 18 SIPs have been developed and approved by Cabinet and the Presidential Infrastructure Coordinating Commission (PICC). The identified projects will provide new infrastructure, assist in terms of rehabilitating and upgrading existing infrastructure, and will also play a crucial role in facilitating the regional integration for African cooperation and economic development on the continent.

These SIPs, particularly those pertaining to the development of rail infrastructure, logistics and industrial corridors, require increasing production of materials such as cement, aluminium, iron and steel, the production of which is inherently carbon-intensive by nature. In the event that technology in these sectors is limited, the costs associated with cleaner technological alternatives are high, the ease of adjusting production processes to new cleaner technologies is low, the possible time frames for such adjustments are high, and therefore emissions would increase drastically in these sectors.

The third example comes from the proposed implementation of the carbon tax. First and foremost, the analysis associated with the carbon tax as presented by National Treasury does not talk to how the tax helps achieve the RBS scenario. Second, the proposal for the levy of the tax ignores the structure and economic reality of different sectors. For example, government, through the IRP, determines the modes of electricity generation or the electricity generation capacity mix. Therefore, a carbon tax will not alter the electricity mix. So the emissions reduction possible from this sector would be limited to that dictated by the IRP. Unless this is changed, the carbon tax is unlikely to result in the reduction of emissions in this sector, which accounts for half of the country's emissions. Moreover, given the structure of the electricity sector, price signals to end-users by way of a carbon tax will not lower the carbon intensity of the electricity fuel mix any further than what the government has already determined through the IRP. Finally, the tax scheme ignores that IRP 2030 uses GHG emissions as a constraint in the underlying modelling that determines the future electricity generation capacity. Specifically, it limits electricity sector emissions to 275Mt per annum from 2025 in line with the PPD trajectory suggested by the LTMS.

The political settlement argument, complemented by the poor coordination between energy planning and carbon tax, explains why the tax would not be effective in changing production or consumption behaviour in the electricity sector. In the absence of intervention in the economic organisation of social power and significant institutional and political retooling, the carbon tax would degenerate into an added cost at all levels in the economy.

The issue at stake here is that the Constitution, architecture and functioning of the intertwined social, economic and political power (enshrined in the political settlement) is not conducive to (at best) and resists (at worst) tax and rent reforms. The MEC does not favour the carbon tax and has been vocal about competitiveness impacts, job losses, and the further weakening of the mining and manufacturing sectors. There has been little discussion of the new economic activities that could be incentivised by the tax, the shift of the manufacturing base in the country to cleaner technologies and products, and investments in the related capabilities. The implementation of the tax has been postponed from 2013 to 2016 and it would not be wrong to say that the objective reality and the bargaining power of the MEC have played a role in this shift in time frame.

PART 3: CLIMATE POLICY – LEARNING AND INSIGHTS FROM ELSEWHERE

In the country's defence, it must be said that government's ambivalent attitude to climate mitigation and the failure to align energy and related policy to the envisaged future mitigation objectives is not unique. Examples of intransigence in addressing climate change can be found in other countries. Also highlighted are instances where governments have been proactive in implementing effective mitigation measures and policies. The countries chosen are those that present an opportunity of learning for South Africa. These include Russia, China, India and Brazil in their capacity as BRICS partners, facing similar trade-off challenges to those of South Africa and including a mix of political regimes and natural resources. As a counterpoint, Chile and Australia are included as examples of two countries that seem to be following divergent pathways to climate mitigation. Included as well is Canada, whose federal model is an interesting example of the challenges posed by internal trade-offs, very similar to the trade-offs required in the UNFCCC process in which jurisdictions of vastly different sizes, economic clout and commitment to climate action pursue their own interests (McCarthy, 2015).

Australia (ranked 60), the worst performing industrial country in the German Watch/CAN Climate Change Performance Index 2015[27] (Burck, 2014), is like South Africa (ranked 37) (ibid.), abundant in low-cost coal and with an energy mix dominated by coal-fired electricity generation and energy-intensive industries that make a significant contribution to economic growth, employment and regional development in Australia (DIICCSRTE, 2013). Though its unconditional Copenhagen pledge involves a five per cent reduction from 2000 levels by 2020 (Climate Action Tracker, 2014), emissions are set to increase substantially under the Australian Government's climate policies to more than 50 per cent above 1990 levels by 2020. The Copenhagen pledge, even if fully achieved, would allow emissions to be 26 per cent above 1990 levels of energy and industry GHGs. In addition, it has reversed legislation, in particular repealing the core elements of their Clean Energy Future Plan, effectively abolishing the carbon-pricing mechanism that sought to reduce their renewable target, and blocking other clean energy and climate policy measures. This in spite of the effectiveness of the carbon pricing mechanism in reducing emissions from the electricity and other covered sectors, reducing these by about seven per cent per annum. Up until the time of repeal, this reduction was projected to have been sufficient to meet Australia's Copenhagen pledge (Climate Action Tracker, 2014).

Then there is Russia (ranked 56) (ibid.), another resource rich country and the third largest emitter of greenhouse gases after China and the USA, with one of the highest per capita energy-related CO_2 emissions in the world (Matthews, 2014). Historically, Russia has made little effort to curb its carbon emissions, and has felt little need to invest in renewable resources because of its enormous natural gas, oil and coal resources (De Micco, *et al.* 2013). Ironically, Russia was one of the biggest benefactors of carbon credits under the mechanisms of the Kyoto Protocol due to the severe downturn of its economy in 1990. The economic recovery subsequently, and accompanying growth, has led to a growth of GHG emissions, but Russia will remain in over-compliance with Kyoto targets without any mitigation measures as their

27. On the basis of standardised criteria, the index evaluates and compares the climate protection performance of 58 countries that together are responsible for more than 90 per cent of global energy-related CO_2 emissions. The revised methodology is still primarily centred on objective indicators. Thereby, 80 per cent of the evaluation is based on indicators of emissions (30 per cent for emissions levels and 30 per cent for recent development of emissions), efficiency (5 per cent level of efficiency and 5 per cent recent development in efficiency), and renewable energy (8 per cent recent development and 2 per cent share of total primary energy supply). The remaining 20 per cent of the CCPI evaluation is based on national and international climate policy assessments by approximately 300 experts from the respective countries (Burck, 2014).

carbon emissions are still lower. As a result, Russia (and other Eastern European countries) is not subject to any real caps or incentives to reduce their emissions because their caps are calculated from a high base of the USSR peak emissions (Lioubimtseva, 2010).

At this stage, nothing indicates that Russia intends to change direction. Russia has refused to join the post-2013 extension of the Kyoto protocol. 'Without the effort and dedication of a small group of committed climate experts and stakeholders, Russia's climate policy would boil down to a simple "green" image-building exercise targeted towards other governments – much talk, very little walk' (Kokorin and Korppoo, 2014). In spite of recently initiated legislation to address climate change, and Prime Minister Dmitry Medvedev's 'willingness' to consider limiting carbon emissions and support for renewable energy investment – which is driven by economic rather than climate change issues – Russia will not deviate much from the economically viable development path that is very close to a BAU trajectory that 'rejects additional costs associated with emission reductions'. To conclude, Russia is driven by economic and geostrategic and political factors rather than issues relating to climate change and is pursuing a climate change strategy that reflects the realities of local energy supplies – whether or not regions and states possess significant reserves – as well as government's pursuit of national security. It is, in other words, impossible to draw a clear line between energy resources, climate policies and geopolitics (De Micco, *et al.* 2013).

Chile, unranked by the Climate Change Performance Index 2015, presents quite a different picture and speaks to countries' pursuit of energy security. It has pledged an emission reduction level of 20 per cent below BAU in 2020 (as projected from 2007), using a mix of measures such as Nationally Appropriate Mitigation Actions (NAMAs). There has also been a change in government's previous strategy to address increasing demand, replacing additional coal-fired power plants with the development of LNG import facilities and investment in oil and gas explorations. The government's Energy Agenda 2014–2018 also proposes additional policies that may contribute to further emissions reductions, and it implemented the Non-conventional Renewable Energy Law, which aims to achieve a 20 per cent renewable energy target in 2025 (Energy Tracker, 2014).

Chile is the fifth-largest consumer of energy in South America, but is only a minor producer of fossil fuels and is heavily dependent on energy imports (EIA, 2014). It is the world's leading copper producer, and many copper mines in the country are powered by coal-fired power plants that rely on imported oil and gas. Importing around four-fifths of its current fossil fuel demand is economically disadvantageous as well as environmentally costly for Chile. Government's view is that the private copper mining companies should use technologies to reduce their pollutants or change the fuel that they use, and a carbon tax will support this (Phillips, 2014).

In this context, the announcement made by President Bachelet in September 2013 that Chile intends to enact new environmental tax legislation, making the country the first in South America to tax carbon dioxide (CO_2) emissions, comes as no surprise. The intention is to start measuring carbon dioxide emissions from thermal power plants in 2017 and start charging the new tax from 2018. It will target the power sector, particularly generators operating thermal plants with capacity equal or larger than 50 MW, while thermal plants fuelled by biomass and smaller installations will be exempt. Four companies are expected to pay the bulk of the new tax: Endesa, AES Gener, Colbún and E.CL., and these companies have warned of price hikes in electricity rates and complained that other industrial sectors were excluded (Teixeira, 2013).

Once again, while the proposed carbon tax points to a desire to mainstream climate change mitigation, there are real economic and geopolitical factors underpinning climate change policies with energy use particularly relevant given its links to economic structure and development. In the Chilean context this is a viable prospect. Chile has vast renewable energy potential and in 2013 it doubled its renewable energy target, requiring utilities to get 20 per cent of their power from renewable sources by 2025. It has extremely high potential for solar power generation and has good wind power potential both on and offshore. It could also develop significant geothermal power by taking advantage of its position along the 'Ring of Fire', a tectonic hot spot. Its renewable energy capacity grew 40 per cent in 2013 to just over one gigawatt, and more than 10 times that capacity of energy has been proposed in other renewable projects, most of which are solar (Phillips, 2014).

Canada, ranked 58th, is probably best known for its decision to withdraw from the Kyoto Protocol in December 2011, only hours after the UN summit was over (Holmes, 2012). However, despite this, Canada has maintained its target under the convention – a 17 per cent reduction in emissions below 2005 levels by 2020, which translates to a seven per cent increase in emissions from 1990 levels. On the face of it, Canada's Intended National Determined Contribution (INDC) submitted in May, and described as 'inadequate', is equivalent to a reduction of two per cent below 1990 industrial GHG emissions. However, the inclusion of land use, land use change and forestry, and net-net accounting is likely to be the equivalent of 11 per cent of 1990 industrial emissions (Climate Action Tracker, 2015).

What underpins the recalcitrance of Canada to commit more wholeheartedly to the climate change mitigation project? Arguably, Canada is facing similar challenges to the UNFCCC where geopolitics and disagreements over allocation of the cost of mitigation between North and South are standing in the way of a binding protocol. Canada follows a federal model that translates into a 'jurisdictional divide on climate change'. While the provinces are responsible for natural resource management, including oil and gas developments and their electricity sectors, the federal government regulates pollution and greenhouse gases. Accordingly, all major sources of emissions contributing to climate change can be addressed through both federal and provincial policies. This has resulted in some provinces taking the lead, whilst others (such as Alberta and Saskatchewan), rather than acting to curb global warming emissions, are investing heavily in more polluting industries, in particular the development of oil sand fields. Although these two provinces only represent 14 per cent of the population and 10 per cent of the economy, their GHG emissions are nearly five times what they are in the rest of the country (Holmes, 2012: 8).

Essentially, the debate is twofold with one camp arguing for:

> *... a level playing field with industries and citizens facing the same rules across the country – could produce a dynamic environment with all provinces bringing their strengths and working together while ensuring each region of the country takes responsibility for its impacts* (Holmes, 2012: 9).

Whilst the other argues that:

> *Canadian climate policies [must] recognize and accommodate differences in regional circumstances, similar to the recognition of national circumstances in the UNFCCC process* (Gibbons, 2012).

Not surprisingly, Alberta and Saskatchewan favour the second option.

To date, the federal government has sidestepped the issue, which explains the uncoordinated, unilateral action taken by each province and the federal government (Gordon and Macdonald, 2011). In allowing provinces to determine their own climate policy, the federal government has managed to avoid setting a more ambitious national target and then meeting it – a difficult task to execute in a country 'where the richest western provinces rely heavily on oil and coal for export and power, and the most populous central Canadian provinces import virtually all their fossil fuels' (McCarthy, 2015]. Essentially, and speaking to other countries highlighted in this section, Canada faced the challenge of formulating and coordinating a climate policy that reconciles the needs and strategies of a Chile with those of a Russia.

South Africa's other BRICS partners, China, Brazil, and India, (ranked 45, 49 and 31 respectively by the Index) (ibid.), have formally pledged under the UNFCCC to a quantified national-level or economy-wide objective to limit the growth of GHG emissions. For China and India this objective is intensity-based, expressed in carbon dioxide (CO_2) emissions per unit of gross domestic product (GDP), with the aim to reduce their economy's CO_2 intensity below the 2005 level by 2020, though in both countries the goal is expressed as a range: a 40–45 per cent reduction in China, and a 20–25 per cent reduction in India. Brazil, like South Africa, has set goals against BAU emissions projections in 2020, i.e. GHG emissions as they are expected to be in the absence of new policy (Moarif and Rastogi, 2014).

In all three countries, climate change is not necessarily the single or principal driver of renewable energy development. In addition, there are suggestions that both China and Brazil implement mitigation actions as long as there are no other interests involved (Ucros, 2014). In instances where conflicts of interest develop with environmental policies, both countries have prioritised development over climate mitigation policies. This is also the case with India, which intends to drive its development by increased imports of coal.

A more country-perspective lens reveals the paradoxical nature of China to date. As the world's largest greenhouse gas emitter, it has a poor reputation on environmental issues and was regarded as obstructive at the Copenhagen climate change talks in 2009. Subsequently, it has invested significant resources into policies that reduce greenhouse gas emissions (Williams, 2014); mandated that state-owned enterprises should purchase renewable electricity (Moarif and Rastogi, 2014); set quantitative and binding energy targets to reduce energy intensity (energy consumption per unit of GDP) by 20 per cent; and set goals for increasing 'non-fossil' – renewable and nuclear – energy to 10 per cent of its primary energy. Nevertheless, China continues to be one of the world's largest producers of coal, which is also its dominant source of energy. Added to this is its commitment to economic growth, which underpins poverty alleviation, social stability and, ultimately, government legitimacy, even if it is not compatible with emission reduction targets.

Like China, India faces the twin challenges of fast tracking its development towards poverty reduction on one hand, and on the other, responding to environmental threats like climate change by avoiding and reducing rising greenhouse gas (GHG) emissions. India pledged to reduce the emission intensity of its GDP by 20 to 25 per cent by 2020 compared to 2005 levels (Carbon Action Tracker, 2014). The country's National Action Plan on Climate Change released in 2008, with eight goals, covers both mitigation and adaptation. It targets an increase in solar power generation as a key focus area to help the country become a global leader in solar energy deployment, and the extension of access to electricity, energy efficiency initiatives to yield savings of 10,000 MW by 2012, a 20 per cent improvement in water use efficiency through pricing, and other measures to mitigate water scarcity that is projected to worsen as a result of climate change, and afforestation of six million hectares of degraded forest lands, and expanding forest cover from 23 per cent to 33 per cent of India's territory.

Subsequently, the government has taken various measures to develop renewable energy and now plans to accelerate the deployment of renewable energy to more than 160 GW by 2022, including 100 GW solar energy and 60 GW wind energy (MNRE 2015). It has launched an energy efficiency scheme based on the Perform, Achieve, and Trade mechanism, whereby the country's largest energy users are set benchmark efficiency levels, with trade of energy savings credits occurring between participants who exceed their targets and those who fail to meet them.

However, the Indian government is clear that an environmental protection programme needs to be calibrated to the country's development needs that include expansion of infrastructure, home-building programmes, affordable energy generation, poverty alleviation, access to education, healthcare, and equal job opportunities (MoP, 2015). The government believes that, for a country whose per-capita GDP is a modest $1,500, it is unreasonable to burden the common person with significantly higher costs in the present, considering the large investments required, while major polluters in the world have not done so themselves during their own development period (MoP, 2015). It has been clear to state that, from a cost perspective, while the trajectory of renewables looks promising, fossil-fuel-based thermal power continues to be the cheapest source of electricity generation. Consequently,

India aims to double the use of domestic coal from 565 million tons in 2013 to more than a billion tons by 2019 (Harris, 2014). The government is trying to sell coal-mining licenses as swiftly as possible after years of delay and has signalled that it may denationalise commercial coal mining to accelerate extraction (Harris, 2014).

Brazil is sitting pretty by comparison, with 80 per cent of its electricity produced by hydropower (Ritti, 2015) and it has been effective in using state-owned companies to establish a competitive biofuels market. However, it is not clear if the emerging biofuel industry has been developed for environmental purposes. Ucros (2014) argues that this industry is an initiative to create jobs – over 20,000 jobs – and to create a stimulus for family agriculture. Brazil has also maximised its oil production rent by channelling funds into preferential financing programmes benefiting climate mitigation activities (Moarif and Rastogi, 2014). It has pledged to reduce its emissions by 36.1 per cent to 38.9 per cent in 2020 compared to BAU emissions. According to the analysis of Climate Action Tracker (2014), the country will meet this pledge with current policies.

Similar to China, Brazil needs to balance development objectives with environmental concerns. Its Brazilian National Plan on Climate Change states unequivocally that 'the government will support and encourage the development of new industries on clean energy and technology as long as it contributes to the country's economic growth' (Ucros, 2011: 20). This tension is clearly demonstrated by the trade-offs between protecting its rain forest and agricultural needs. Much of Brazil's climate policies focus on combating carbon emissions caused by deforestation, the largest source by far in Brazil (Ucrós, 2011). The central pieces of policy are the national Forest Code, the Action Plan for Deforestation Prevention and Control in the Legal Amazon (PPCDAm), and the Cerrado (PPCerrado) (Carbon Action Tracker, 2014). However, controlling deforestation is challenging. This is not only because much of it is done illegally, but also because of Brazil developing into an agribusiness superpower, which has led to an increase in demand for farmland. Increased soybean production for the development of biofuel has further exacerbated the situation (Ucros, 2011).

However, Brazil's present energy mix might change dramatically in the near future, which could impact considerably on its ability to fulfil its pledges. Hydropower has proven to be vulnerable to droughts. Its biofuel industry is under pressure since the discovery of offshore oil reserves in 2007. Government has contributed $60 billion of taxpayers' money toward pumping up Petrobras and it has subsidised gasoline and diesel, impacting negatively on the biofuel industry. Similarly, fossil fuels are seen as an alternative to hydropower, and 70 per cent of the new energy investments in Brazil over the next decade, some $300 billion worth, are projected to go to fossil fuel projects (Ritti, 2015).

PART 4: RECOMMENDATIONS AND CONCLUSION

The objectives for the LTMS were for South African stakeholders to understand future scenarios for climate action based on the best available information and provide the base for the development of long-term national climate policy as well as positions for international negotiations. However, the LTMS and the policies it inspired have not been wholly effective.

Given the complexity of the challenges highlighted in this chapter, coupled with the path dependency of the MEC, and the maintaining of coal power plants, albeit with a 'modest contribution' from renewable energy (Satgar, 2014),[28] the question is whether policies can break the status quo and lead to real change. Yet, this situation is not unique. The experience from other countries shows that policies rarely address climate change in isolation; rather, they are designed to fulfil a range of parallel objectives, be they energy security, reduced air pollution, economic restructuring, or targeted industrial development (Moarif and Rastogi, 2014). Access to a ready supply of fossil fuels is a major hindrance in the sustained roll-out of climate policy, as shown by Russia, Australia, and the backtracking of Brazil and Canada.

What does this mean for South Africa? The country does not lack policies, strategies or climate-related goals. The big bottleneck is the implementation of policy interventions and sector strategies that are burdening not just the sectors covered by policies, but the country at large through adverse multiplier effects. What is required are policy synergies that integrate the social and economic priorities of the country, across and within government, specifically between policies for climate change, industrial development, job creation, poverty reduction and the energy sector.

The government needs to exploit critical junctures such as the present electricity crisis to move beyond short- and medium-term vested interests, as well as old policies and understandings, and implement solutions that deliver results. Critical junctures are moments when existing political and

28. This is also spelt out in the NDP – 2030, as follows: 'South Africa has a rich endowment of natural resources and natural deposits, which, if responsibly used, can fund the transition to a low-carbon future and a more diverse and inclusive economy' (South African Government, 2012 in Satgar, 2014: 143).

institutional structures fail to provide either adequate solutions to pressing problems or explanations of challenging events, and thus lose governance legitimacy and their ability to determine action and interpretation (Katznelson in Sorenson, 2015).

In critical junctures, the 'structural constraints imposed on actors during the path-dependant phase are substantially relaxed'. This provides an opportunity for 'policy entrepreneurs and other actors to reshape existing institutions and create new arrangements'. It provides the possibility of multiple futures, a 'window of political opportunity', albeit still constrained by larger economic, political and ideological structures and forces. However, actors can influence outcomes and, where there are 'positive feedback effects', break the chain of path dependency and in so doing change the rules (Sorenson, 2015). If the new rules become long lasting, they have the potential to change the existing 'patterns of distributional advantage' (Lowndes in Sorenson, 2015).

Recalling Milton Friedman, the politically inevitable is upon us. The transition from the politically impossible to politically inevitable compels us to re-imagine and rethink our policy 'utopia' refracted through Realpolitik (the dystopia of the working and functioning of the political settlement) and grounded in material re-making of economics and development. Reconnecting real world economics and transformative mitigation is the imperative.

References

Baker, L. 2013. 'Low Carbon transitions in South Africa's minerals-energy complex?' Paper presented at the International Initiative promoting political economy, International Social Sciences, The Hague.

Burck, J., Marten, F. and Bals, C. 2014. 'The Climate Change Performance Index Results: 2015'. Available at: http://www.legambiente.it/sites/default/files/docs/germanwatch ccpi2015results8.12.14.pdf [Accessed 8 June 2015].

Climate ActionTracker. 2014. 'Australia: Emissions set to soar by 2020', December. Available at: http://climateactiontracker.org/news/179/Australia-Emissions-set-to-soar-by-2020.html [Accessed 10 January 2015].

Climate Action Tracker. 2014. Available at: http://climateactiontracker.org/ [Accessed 30 October 2014].

Cook, S., Smith, K. and Utting, P. 2012. 'Green Economy or Green Society? Contestation and Policies for a Fair Transition', Occasional Paper Ten, November. Available at: http://www.unrisd.org/80256B3C005BCCF9/httpNetITFramePDF?ReadForm&parentunid=B24EA25289BD528AC1257AC5005F6CA5&parentdoctype=paper&netitpath=80256B3C005BCCF9/(httpAuxPages)/B24EA25289BD528AC1257AC5005F6CA5/$file/10%20Cook-Smith-Utting.pdf [Accessed: 11 October 2014].

Cornwall, A. and Brock, K. 2005. 'What do Buzzwords do for Development Policy? A Critical Look at "Participation", "Empowerment" and "Poverty Reduction"', *Third World Quarterly*, Vol. 26, No. 7, pp. 1043–1060.

Death, C. 2014. 'The Green Economy in South Africa: Global Discourses and Local Politics', *Politikon: South African Journal of Political Studies*, 41:1, 1–22, DOI: 10.1080/02589346.2014.885668 Available at: http://dx.doi.org/10.1080/02589346.2014.885668 [Accessed: 11 December 2015].

De Micco, P., *et al.* 2013. 'Policy Briefing on EU and Russian policies on energy and climate change', *Policy Briefing*, December. Requested by the European Parliament's Delegation to the EU-Russia Parliamentary Cooperation Committee. Available at: http://www.europarl.europa.eu/RegData/etudes/briefing_note/join/2013/522304/EXPO-AFET_SP(2013)522304_EN.pdf [Accessed: 4 April 2015].

Department of Energy, Republic of South Africa. 2013. Integrated Resource Plan for Electricity (IRP) 2010–2030, Update Report 2013.

Department of Industry, Innovation, Climate Change, Science, Research and Tertiary Education (DIICCSRTE)© Commonwealth of Australia. 2013. 'National Communication on Climate Change. A report under the United Nations Framework Convention on Climate Change, Australia's Sixth', Available at: http://unfccc.int/resource/docs/natc/aus_nc6.pdf [Accessed: 20 January 2015].

Fakir, S. and Gulati, M. 2012. 'Carbon Tax in South Africa'. Presentation to Economic Development Department and National Treasury, 18 May.

Gibbens, R. 2009. 'Sharing the Load. Addressing the Regional Economic Effects of Canadian Climate Policy: A Critique of the Pembina Institute and the David Suzuki Foundation Study of the Economic Implications of Reducing Greenhouse Gas Emissions in Canada'. *The West in Canada Research Series*, December. Available at: http://cwf.ca/pdf-docs/publications/sharing-the-load-december10-2009-CWF-critique-of-pembina-suzuki-report.pdf [Accessed 8 June 2015].

Gordon, D. and Macdonald, D. 2011. 'Intergovernmental Coordination in Australia and Canada'. Available at: http://www.cpsa-acsp.ca/papers-2011/Gordon-Macdonald.pdf [Accessed 8 June 2015].

Harris, G. 2014. 'Coal Rush in India Could Tip Balance on Climate Change', *The New York Times*, November 17. Available at: http://www.nytimes.com/2014/11/18/world/coal-rush-in-india-could-tip-balance-on-climate-change.html?_r=0 [Accessed 14 February].

Held, D. and Hervy, A. F. 2009. 'Democracy, Climate Change and Global Governance', *Policy Network* paper, November.

Holmes, M. 2012. 'All Over the Map 2012: Institutions and Federal Climate Change Governance: a Comparison of Provincial Climate Change Plans'. David Suzuki Foundation, March. Available at: http://www.davidsuzuki.org/publications/

downloads/2012/All%20Over%20the%20Map%202012.pdf [Accessed 8 June 2015].

Khan, F. and Mohamed, S. 2014. 'From the MEC to the Green Economy Conglomerate: Wiring a Green Economy-Green Society-Developmental State Circuit', *New Agenda: South African Journal of Social and Economic Policy* 56: 47–53.

Khan, M. 2013. 'Political Settlements and the Design of Technology Policy.' In: Stiglitz, J. and Lin, J. Y. and Patel, E. (eds.), *The Industrial Policy Revolution II. Africa in the 21st Century*. London: Palgrave, pp. 243–280.

Kokorin, A. and Korppoo, A. 2014. 'Russia's Greenhouse Gas Target 2020: Projections, Trends, and Risks', Friedrich-Ebert-Stiftung, April. Available at: http://library.fes.de/pdf-files/id-moe/10632.pdf [Accessed: 15 April 2015].

Lioubimtseva, E. 2010. 'Russia's Role In The Post-2012 Climate Change Policy: Key Contradictions And Uncertainties', *Forum on Public Policy*. Available at: http://forumonpublicpolicy.com/spring2010.vol2010/spring2010archive/lioubimtseva.pdf [Accessed: 30 March 2015].

Marquard, A., Trollip H. and Winkler, H. 2011. *Opportunities for and Costs of Mitigation in South African Economy*. Republic of South Africa: Energy Research Centre, University of Cape Town and Department of Environmental Affairs.

Matthews, H. D., *et al.* 2014. 'National Contributions to Observed Global Warming', Environmental Research Letters, Vol. 9, no. 1. Available at: http://iopscience.iop.org/1748-9326/9/1/014010/pdf/1748-9326_9_1_014010.pdf [Accessed 20 May 2015].

McCarthy, S. 2015. 'Greenhouse Gas Emissions: Who's Responsible for Climate Policy in Canada?' *The Globe and Mail*, March 23. Available at: http://www.theglobeandmail.com/news/politics/canadas-provinces-are-taking-the-the-lead-on-climate-but-should-they/article23583907/ [Accessed 9 June 2015].

Ministry of New and Renewable Energy, India. 2015. 'PM To Inaugurate First Renewable Energy Global Investors Meet On Sunday; Over 2500 Delegates and Exhibitors from 45 Countries To Participate In The Three Day Event; Piyush Goyal Says RE-INVEST Aims To Project India As An Attractive Investment Destination And To Evolve Innovative Financial Models'. Press release dated 13 February. Available at: http://pib.nic.in/newsite/pmreleases.aspx?mincode=28 [Accessed 14 February 2015].

Ministry of Power. 2015. 'Modi Government is Committed to Provide Affordable and Clean Power 24x7 For All'. Press release dated 12 January 2015. Release ID: 114557. Available at: http://pib.nic.in/newsite/pmreleases.aspx?mincode=52 [Accessed 14 February 2015].

Moarif, S. and Rastogi, N. P. 2012. 'Market Based Climate Mitigation Policies in Emerging Economies'. Centre for Climate and Energy Solutions, December, Available at: http://www.c2es.org/docUploads/market-based-climate-mitigation-policies-emerging-economies.pdf [Accessed: 20 April 2015].

National Treasury, Republic of South Africa. 2014. 'CARBON OFFSETS PAPER. PUBLISHED FOR PUBLIC COMMENT'. Pretoria.

Phillips, A. 2014. 'Chile plans to enact the first carbon tax in South America', April. Available at: http://thinkprogress.org/climate/2014/04/08/3424238/chile-carbon-tax/ [Accessed 20 March 2015].

Ritti, C. 2015. 'Could the oil drill kill Brazil?' *Brazilian Climate Observatory*, 22 January. Available at: http://www.trust.org/item/20150122112757-n3fyt/ [Accessed 15 March 2015].

Satgar, V. 2014. 'South Africa's emergent "Green Developmental State"?' In *The End of the Developmental State*, Michelle Williams (ed.), University of KwaZulu-Natal Press, p. 126–153.

Scholvin, S. 2014. 'South Africa's Energy Policy: Constrained by Nature and Path Dependency', *Journal of Southern African Studies*, 40: 1, 185–202, DOI: 10.1080/03057070.2014.889361. Available at: http://dx.doi.org/10.1080/03057070.2014.889361 [Accessed 10 February 2015].

Seguino, S. 2014. 'How economies grow: Alice Amsden and the real world economics of late industrialization', Paper presented at the Alice H. Amsden Memorial Lecture, University of the Witwatersrand, Johannesburg, South Africa, 4 September.

Sorensen, A. 2015. 'Taking path dependence seriously: an historical institutionalist research agenda in planning history', *Planning Perspectives*, 30: 1, 17–38, DOI: 10.1080/02665433.2013.874299. Available at: http://dx.doi.org/10.1080/02665433.2013.874299 [Accessed 28 January 2015].

Texeira, M. 2014. 'Chile becomes the first South American country to tax carbon', *Reuters*, September. Available at: http://uk.reuters.com/article/2014/09/27/carbon-chile-tax-idUKL6N0RR4V720140927 [Accessed 20 May 2015].

The Guardian. 2013. 'Japan under fire for scaling back plans to cut greenhouse gases'. Available at: http://www.theguardian.com/global-development/2013/nov/15/japan-scaling-back-cut-greenhouse-gases [Accessed 15 August 2015].

Therborn, G. 2014. 'New masses?' *New Left Review 85*, Jan./Feb., p. 7–16.

Ucrós, C. M. 2012. *Climate Change Mitigation Policies in Developmental States: A Comparative Approach to China and Brazil*, London School of Economics and Political Science MSc: China in Comparative Perspective. Available at: http://www.banrepcultural.org/sites/default/files/munoz_camila_tesis.pdf [Accessed 15 January 2105].

'United Nations Framework Convention on Climate Change'. Undated. *Parties and Observers*. Available at: http://unfccc.int/parties_and_observers/items/2704.php [Accessed 17 June 2015].

US Energy and Information Administration (EIA). 2014. 'Overview of Chile', July.

Available at: http://www.eia.gov/countries/country-data.cfm?fips=CI [Accessed 15 January 2015].

Wainwright, J. and Mann, G. 2013. 'Climate Leviathan', *Antipode*, Vol. 43(1).

Williams, L. 2014. 'China's Climate Change Policies: Actors and Drivers', Lowy Institute for International Policy, July. Available at: http://www.lowyinstitute.org/publications/chinas-climate-change-policies-actors-and-drivers [Accessed 15 January 2015].

CHAPTER 3

Historical Review of the Relationship Between Energy, Mining and the South African Economy

Marie Blanche Ting

ABSTRACT

Mining, energy and industrial development have been key to growth and development in South Africa. It is estimated that South Africa's mineral wealth value is around $2.5 trillion (GCIS, 2012). Mining and the electricity industry co-evolved together due to the abundance of inexpensive coal. The electricity industry was spurred by the needs of a flourishing mining industry around the beginning of the twentieth century. South Africa's electricity industry is embedded in what is known as the Minerals-Energy Complex (MEC). The MEC is described as the relationship between mining, energy-intensive mineral processing, the coal-to-electricity sectors, and parts of the supportive transport and logistics infrastructure (Fine and Rustomjee, 1996). However, the continued reliance of South Africa's electricity system on coal is increasingly under threat from international pressure to mitigate climate change. Additionally, sustained dependence on a resource-based economy is risky, not only because of finite supply but, more importantly, because emphasis on a global commodity market renders the economy vulnerable to short-term volatile capital flows, and less long-term labour absorbing economic growth. The main objective of this chapter is to discuss how South Africa faces the challenges and demands towards sustainable transitions.

This chapter first provides a brief overview of South Africa's mining and energy and the need for sustainable development. After which, the underlying theoretical derivatives and framework are discussed. Subsequently, it centres its discussion on interest, ideas, and institutions. It will analyse how policies, discursive narratives and vested interests related to mining and energy have previously interacted and currently interact, effectively creating barriers for the innovation of more sustainable technologies.

As a way forward, a discussion takes place on a proposed implementation for sustainable mining and energy that includes: the necessary requirements to adequately capacitate the socio-technical systems that govern and implement mining and energy developments; continuous monitoring and evaluation once policies are implemented; understanding the concept and implications of carbon lock-in given the long-term life span of energy expansion; and meta-coordination of strategy implementation. Lastly, a recommendation on further research is given on shifting from a resource-based economy towards a knowledge-based economy. South Africa needs to go beyond beneficiation. Rather, it should expand the process underlying the knowledge in obtaining these minerals and use it to diversify or enhance other sectors of the economy. These could include technological innovation in the mechanisation process, machineries, parts, tools and equipment – all of which could spill over into important sectors such as transportation, information technology, construction, and manufacturing.

INTRODUCTION

South Africa initially built its economy on two major primary sectors: mining and agriculture. Mining has been the foundation of South Africa's industrialisation and later the financialisation of the economy. Over time, mining, energy and industrial development have been key to growth and development in South Africa (Gentle, 2009; Winkler, 2009 and McDonald, 2009). It is estimated that South Africa's mineral wealth value is around $2.5 trillion (GCIS, 2012). A term to describe a tight configuration of industries associated with mining is often called the Mineral Energy Complex (MEC) (Fine and Rustomjee, 1996). The MEC is described as the relationship between mining, energy-intensive mineral processing, the coal-to-electricity sectors and parts of the supportive transport and logistics infrastructure (Fine and Rustomjee, 1996). Despite the recent years of poor performance of the mining sector, mining continues to be an important part of the South African economy and this is unlikely to change for a few decades. South Africa has many strategic minerals of global importance.

South Africa's mining history and its relevance to its economy can be traced as far back as the discovery of diamonds in Kimberley in 1871 and the Witwatersrand main gold reef in 1886. The wealth of its mineral resources is evident with one of the largest reserves of Platinum Group Metals (PGMs), chromium, gold, manganese (high grade ore), alumino-silicates, vanadium and other minerals in the world (Table 1). In addition, the mining sector has also been a source of other competencies such as engineering, technical and production expertise, and broad research and development activities. Most of South Africa's mineral wealth is exported as ores, concentrates, alloys or metals with some degree of beneficiation downstream. The exceptions are iron and steel, as well as polymers from oil and coal, which are used as inputs into the manufacturing sectors (ANC SIMS, 2012). As depicted in Table 1, the mineral wealth reserves in the country have a long lifespan and are expected to continue to play an important role in the country.

Mining and Energy

Mining in South Africa and its associated sectors includes iron and steel, chemicals, non-ferrous metals, non-metallic minerals, electricity and transport. The mining industry is also important to manufacturing as it is linked via activities such as minerals beneficiation and metals production (Winkler, 2009) as well as backward linkages in mining equipment. Therefore, the contribution of the MEC to the economy cannot be viewed in isolation nor can it measured in absolute terms. The largest energy-intensive industrial subsector in South Africa is iron and steel, which consumes around 27 per cent of the total energy used by the industry sector (DoE, 2010). The South African mining and electricity industries co-evolved together due to the availability and abundance of inexpensive coal. The electricity industry was spurred by the needs of a flourishing mining industry around the beginning of the twentieth century (Gentle, 2009 and Winkler, 2009). The major challenge for sustainable development in the context of climate change mitigation lies in the energy sector, which accounts for more than 70 per cent of greenhouse gas (GHG) emissions. The share of final energy demand can be distributed as follows: industry (38 per cent), residential (18 per cent), and transport (28 per cent) and other minor contributions (DoE, 2012). Underlying the energy sector is an intensive coal-

Table 1: A summary of South Africa's mineral reserves and their relative global production percentages

Mineral	Unit	Reserves			Production 2009			Life
		Mass	% World	Rank	Mass	% World	Rank	Years
Alumino-silicates	Mt	51	-	-	0.265	60.2	1	192
Antimony	kt	350	17	3	3	1.6	3	117
Chromium Ore	Mt	5500	72	1	6.762	-	1	813
Coal	Mt	30408	7	6	250.6	3.6	7	121
Copper	Mt	13	2	6	0.089	-	-	146
Fluorspar	Mt	80	17	2	0.18	3.5	5	444
Gold	t	6000	13	1	197	7.8	5	30
Iron ore	Mt	1500	0.8	13	55.4	3.5	6	27
Iron ore including (BC)	*Mt*	*25000*	*10*	-	*55.4*	*3.5*	*6*	*451*
Lead	kt	3000	2	6	49	1.2	10	61
Manganese Ore	Mt	4000	80	1	4.576	17.1	2	874
Nickel	Mt	3.7	5.2	8	0.0346	2.4	12	107
PGMs	t	70000	88	1	271	59	1	258
Phosphate rock	Mt	2500	5.3	4	2.237	1.4	11	1118
Titanium minerals	Mt	71	10	2	1.1	19	2	65
Titanium including (BC)	*Mt*	*400*	*65*	*1*	*1.1*	*18*	*2*	*364*
Uranium	kt	435	8	4	0.623	1.3	10	698
Vanadium	kt	12000	32	2	11.6	25.4	1	1034
Vermiculite	Mt	80	40	2	0.1943	35	1	412
Zinc	Mt	15	3	8	0.029	0.2	25	517
Zirconium	Mt	14	25	2	0.395	32	2	35

(ANC Sims, 2012)

based system, the result of which is nine tons per capita (tpc) of carbon emissions, rivalling that of Organisation for Economic Cooperation and Development (OECD) countries with an average of 11 tpc (World Bank,

2010). Thus, a case for sustainable energy development in the country is relevant because it is considered to be 'carbon intensive'. The continued reliance of South Africa's electricity system on coal is increasingly under threat from international pressure to mitigate climate change. Furthermore, resource-based economies are subject to finite supply and to global commodity cycles that are increasingly cognisant of trends towards environmental sustainability.

The main objective of this chapter is to provide insights into the barriers in South Africa's mining and energy complex that have to be addressed in order for sustainable development to be feasible.

Framework

The analysis used for investigating the feasibility of sustainable energy development in the mining sector was structured around an interests, ideas, and institutions framework (Figure 1). The main theoretical derivative for this framework is drawn from Giddens' (1984) structuration theory. Giddens (1984) defines structuration theory as a means of understanding human social behaviour by analysing the interactions between agency and structure. Agency is the capacity of actors who operate in a defined set of rules and resources that are embedded in structures. According to Geels and Schot (2010), rules are described as cognitive (belief systems, guiding principles, goals and search heuristics), regulative (standards, laws and regulations) and normative (values and behavioural norms), (Geels and Schot, 2010). It is assumed that actors are embedded in these structures, which are both a medium and outcome of action. For the purposes of this chapter, rules and resources are framed according to three areas, namely ideas, institutions and interests.

Ideas refer to meaning and signification that are realised in practice. More specifically, they focus their analysis on discursive narratives, ideologies, belief systems, guiding principles and mindsets that are related to mining and energy in the country. Empirically, this chapter discusses the structure of the economy related to mining and the role it may play in strategy or policy formulation around sustainable development; historical availability of cheap coal and its relevance to trade and competitiveness; the recent trends in the uncertainty around the oil price; the Long-term Mitigation Strategy (LTMS), which has conscientised the efforts needed to lower carbon emissions; and the role of labour and trade unions, which have engaged in important debates in these and related issues.

Figure 1: Summary of three issues, namely: interests, institutions, and ideas, which this chapter seeks to analyse relative to energy, mining and the South African economy

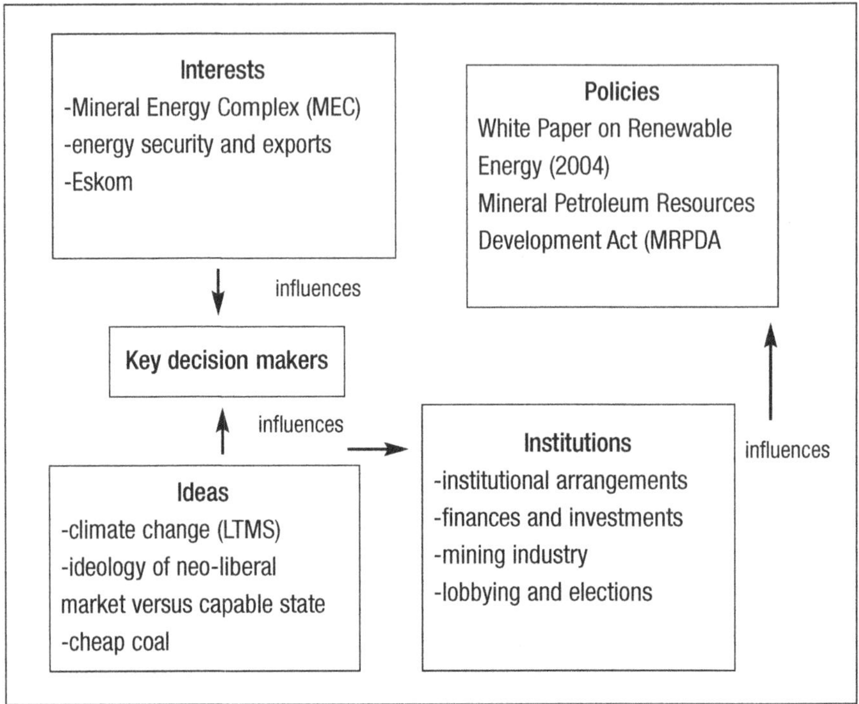

In terms of institutions, these are socially embedded systems of rules through legislation and regulation; not to be confused with organisations that are defined as groups of firms or networks of actors that are assembled together by a collective purpose (Hodgson, 2006). To put it another way, institutions are rules of the game and organisations are players that abide by these rules within a system. It is the interaction between institutions and organisations that configures a socio-economic system. Relevant for this chapter, institutions look at the legislative mandate governing the mining sector that enables the rules of the game; financialisation that serves to strengthen profits and favour the MEC; the role of lobbying and the influence of financial contributions to political parties and, lastly, the key Integrated Resource Plan (IRP) that determines energy decisions in the country.

The last section is on interests and this refers to domination and how power is exercised. The aim is to investigate influential groups and the power they may have in lobbying issues relative to their agendas. These groups include the powerful MEC sectors, the influential role of the state-owned power utility, Eskom, and continued reliance on coal for energy security.

By discussing these three areas, it is hoped that a holistic argument can be made for clearing the stumbling blocks for sustainable development in mining and energy in South Africa.

IDEAS

South Africa's mineral wealth is due mostly to its platinum reserves, followed by coal, palladium, gold, titanium, copper and others. The biggest mining companies in South Africa have significant investments within the Johannesburg Stock Exchange (JSE) and these also belong to an Energy Intensive Users Group (EIUG), which collectively (including associated industries such as materials beneficiation and materials manufacturing) account for 44 per cent of electricity demand in the country (EIUG, 2015). Furthermore, mining provides critical foreign exchange via trade exports; it contributes around 20 per cent to the GDP (direct and indirect) and is an important source of jobs in the country. It is estimated that a social multiplier for mining is a ratio of 10 to 1, thus, given the employment of 1.3 million people, the totality of people dependent on the mining industry can be as high as 13 million. The mining sector is built upon a model of high labour and low wages. However, to what extent this will continue in the future remains unclear, not only because of increasing trends towards mechanisation, but also because of instability within the labour force and wildcat strikes.

Furthermore, the interlinkages of growing inequality and social unrest cannot be ignored because most miners earn wages that are not adequate to support large families. However, this labour model is historically derived and embedded within the socio-economic framework. It should be appreciated that sustainable energy development is only a part of a complex set of problems that policymakers need to address. The country is experiencing developmental challenges that include widening inequality (Gini-coefficient of 0.6), an unemployment rate of 25 per cent, and slow real economic growth of less than 2 per cent GDP per year. Sustainable energy development is part of a much wider set of issues facing the country. To move forward it is important for key decision-makers to realise that some of the underlying issues are rooted within ideologies that constrain advancement. The following sections will now discuss key ideological issues.

Coal as the Dominant Resource

Coal accounts for 70 to 75 per cent of primary energy supply, 93 per cent of electricity generation and 30 per cent of petroleum liquid fuels (Eberhard, 2011; Winkler and Marquard, 2009; Winkler, 2009; Davidson, 2006). Coal production and consumption are intricate parts of the economy. The revenues are derived from a range of activities from export, mining, to the manufacture of synthetic fuels. South Africa has rich coal reserves, albeit low in quality (high ash content). The low-grade coal has an average of 4,500 kcal per kg, ash content of 29.5 per cent and sulphur of 0.8 per cent (Eberhard, 2011). The approximated coal reserve stands at 32 billion tonnes making it the world's largest recoverable deposit (World Energy Council, 2010). South Africa's energy landscape has historically been tied to the strategic needs defined by the government at specific times. Therefore, between 1948 and 1994, the main imperative from the government was energy independence and self-sufficiency (particularly as it grew isolated from the international community due to sanctions and other forms of isolation). Thus, the apartheid government decided to secure its energy supply by utilising the coal to liquids (CTL) process for fuel production, meaning that large quantities of both electricity and fuel are produced from coal: the former is provided by Eskom and the latter by Sasol. Sasol is the world leader in the processing of coal to synthetic fuels, having started production in the 1950s. It produces more than 40 per cent of South Africa's domestic fuel and contributes four to five per cent of the country's GDP. Coal accounts for more than 90 per cent of the electricity produced in the country; the rest comes from nuclear power stations (five per cent), hydro and pumped storage (four per cent), as well as biomass, solar, and wind, which together account for less than 1 per cent (Heun *et al.*, 2010; Pegels, 2010; Davidson, 2006; Winkler, 2005).

Sasol and Eskom have a combined carbon dioxide emission of 300 million tonnes (Mt), which accounts for 67 to 69 per cent of the total GHG output (Raubenheimer, 2011; Sasol 2010; UNSD, 2008). Coal is an entrenched regime with powerful industrial players that have a huge role to play in the country's economy. To what extent the country reduces its reliance on coal is certainly a central issue in terms of barriers to sustainable development. It is this balance between ensuring energy security through its rich coal reserves, but also through simultaneous sustainable practices, which require the least economic trade-offs. However, choosing carbon-intensive pathways that continue to lock in and reinforce business as usual practices may in fact undermine desired developmental goals. Sustainable development can offer ways of diversifying the economy in a manner that is in line with global trends towards greener economies.

The Challenge of Cheap Coal and Electricity

The availability of inexpensive coal was critical to the establishment of the mining and electricity industries. The term 'mineral energy complex' (MEC) was developed by the seminal study called *The Political Economy of South Africa*, by Fine and Rustomjee, in 1996. The MEC has fundamentally formed the energy economy of the country, which in turn has a major role in the energy landscape of the country. The close involvement of the MEC has resulted in the country having one of the lowest electricity prices in the world. It is reported that South Africa's electricity prices were consistently 40 per cent of average United States prices for the last four decades (Winkler and Marquard, 2009). In fact, Eskom, through its association with the MEC, has taken advantage of large economies of scale in coal mining and power generation. Thus, guaranteed demand meant that favourable contracts could be negotiated over the long term. Furthermore, most of Eskom's assets relating to electricity generation had been paid off, which means that its expenditure was limited to low running costs. Although South Africa has known since 1997 – as shown in its White Paper on Energy – that an electricity deficit would be reached by 2007, no major new plants had been commissioned by this time. Thus, with an ageing electricity infrastructure and lack of maintenance due to low generation reserves, the country has undergone a series of blackouts since 2008. This means that, for a period of seven years, the country has been experiencing uncertain electricity supply. To mitigate electricity shortfalls, diesel generators have been used at significant cost. Unfortunately, the country's state-owned power utility has recently faced a series of credit downgrades, and the cost of using diesel has only aggravated this situation and has serious implications for its ability to borrow money for new energy builds.

Cheap coal and cheap electricity have been central to South Africa's competitiveness. Specific industrial policy has promoted electricity-intensive investments such as mineral processing (Winkler and Marquard, 2009). Moreover, special electricity supply contracts between, for instance, Eskom and BHP Billiton and its aluminium smelters were revealed in 1992, and again in 2001, showing sales at below cost prices. The special deal had effectively tied Eskom to supplying BHP Billiton with cheap electricity for more than 20 years. To this day, these special deals continue to stress Eskom's finances. However, new electricity builds require financing and the reality of cheap electricity cannot continue. Any adjustments to the dominant paradigm in the electricity sector are bound to have an impact on energy prices, as well as the labour force, and disrupt a huge source of investment and income for the country. Cheap electricity can no longer be considered for new plants. The present new build programme, irrespective of the sources (coal, nuclear or renewables), has to factor in current levelised costs.

Coal and its Relevance to Civil Energy and Households

It is important to note that coal not only provides the country with significant liquid fuels but it is also a means for government to provide basic energy access. 'Civil energy' has been described as a social development goal by providing energy to households, commerce and services (Winkler and Marquard, 2009: 51). Energy is crucial for lighting, cooking, heating, clean water and sanitation, health care, transport, industries and telecommunication. A strong narrative of the post-apartheid government was to address the imbalances of the past by ensuring basic services such as 'electricity for all'. Almost two-thirds of the South African population did not have access to basic electricity before 1994. The rapid electrification programme resulted in electrification rates that grew from 30 per cent in 1990 to 73 per cent in 2006 (Winkler and Marquard, 2009). Household electricity access and final energy residential demand should not be confused. As referred to earlier, in the context of overall energy demand, the share of residential consumption stands around 15 per cent, but electricity access to the national grid is currently around 70 to 80 per cent.

Thus, post-1994, South Africa had a period where the available electricity infrastructure did not match the rapid growth in demand. At present, the country suffers from a double 'lock-in' of a high carbon pathway due to the dominance of the MEC, but also its continued reliance on coal for its civil energy. Civil energy is a strong principle for government because it is a developmental objective for energy access. Historically, this meant access to grid connection with coal as the main resource. This carbon lock-in makes it difficult to address issues of sustainable energy development because there are sunken investments, entrenched regimes, and historical vested interests. Thus, in an energy policy setting, it can be appreciated that a complexity of issues has to be carefully considered.

Influence of Labour: The Role of Trade Unions

Carbon cycles do not follow political cycles, in that climate change requires a long-term vision. Its effects are anticipated in the medium term, which, though beyond the duration of most politicians' careers, requires immediate action. Hence, the certainty of mitigating options such as renewable energy needs strong arguments so as to achieve buy-in from sectors that will ensure their implementation. In this respect, one has to factor in the influence of members of society and their awareness of the effects and mitigation of climate change. Moreover, people's sentiments equal votes, which cannot be underestimated given that this is the lifeblood of politics.

> *... while voters tend to be strongly supportive of the idea of compliance with the international environmental treaties, they can simultaneously be strongly resistant to the reality of higher taxes and energy prices ... the political incentives thus can be very different between ratification and implementation of that international commitment ... Harrison and Sundstrom* (2007: 15).

In South Africa, a sector that is influential and important to the vote is that of the trade unions.

> *Attending to the issues raised by them will constitute the surest route to electoral success* (Stedman, 1994: 191).

Through the Tripartite Alliance, it is often the case that the South African Communist Party (SACP) and Congress of South African Trade Unions (Cosatu) field their preferred candidates through the African National Congress (ANC), and consequently hold senior positions in the ANC as well as influence party policy and dialogue (Johnson, 2011). There are influential people who hold various positions in one or more of the parties that constitute this alliance. Added to this complexity is that the trade unions emerged mostly from the minerals and mining sectors. Hence, their significance cannot be underestimated since it permeates both the political and economic spheres. In addition, hundreds of prominent unionists now play leading roles in institutions of the new South Africa and are able to influence the broader processes of change that are unfolding. Thus, a formidable group of people from politicians, trade unionists, to business people are all interlinked with each other.

The effects of continued mining strikes in the country are indicative of how intricately the mining sector is embedded in the economy. The five-month platinum strike from January to June 2014 resulted in economic contraction during the first quarter of that year by as much as 0.6 per cent. Mining strikes had an effect on manufacturing as well as industries that included petroleum, basic chemicals and iron ore (Singh, 2014). In addition, tensions within Cosatu and its affiliates have resulted in the exit of the National Union of Metal Workers of South Africa (Numsa), which accounts for an estimated 17 per cent of Cosatu's membership (Letsoalo, *et al.* 2014). Part of the tension in this regard derives from contestation over economic policy, and this cannot be taken lightly. It can lead to policy formulation that is logical, but often hampered at the implementation stages due to the tensions among various actors. Evidence of this is the National Development Plan (NDP), which has ambitious, bold plans for the country and yet seems to be hampered at the level of implementation due, in part, to these tensions. As long as these ideological mismatches are not addressed, they will continue to be a source of policy incoherence and implementation inertia.

Uncertain Oil Price

In a world where the oil price has been assumed to take an upward trajectory, energy investments are decided accordingly. These pertain to new projects in oil and gas exploration and production, as well as renewable energy technologies. Often these investments are decided on the basis of economic costs particularly associated with the price of crude oil. However, the basis of these assumptions no longer holds, given that between June 2014 and January 2015 the oil price dropped by 60 per cent. Thus, associating energy investments with the prediction of an oil price is becoming increasingly uncertain. Added to this are unpredictable geopolitics, global economic developments and competing alternative energy options. Renewable energies are also increasingly on par in terms of economic costs in comparison to new fossil fuel derived plants. A good example is the cost of some solar technologies becoming competitive with natural gas. Again, these expectations may change as the oil price fluctuates. The main point is uncertainty, particularly since there are now more competing energy sources available. Conversely, it has been noted that, in spite of these uncertainties, the development of flex/fuel engines can provide ways to bypass such ambiguities. Again, under these conditions an important balance is needed in choosing the optimal technology that caters to societal needs. South Africa does have huge amounts of coal reserves; potential shale gas in the Karoo region; and the Northern Cape is an ideal place for renewable solar plants. Moreover, there are examples of a hybrid solar Photo Voltaic (PV) cell usage combined with diesel engines at a scale of 1 MW being used in Chile and South Africa (Lazenby, 2014; Chadbourne, 2014). Perhaps a modular approach of hybrid engines comprising solar energy with gas could be applicable to South Africa given its historical transmission configuration and natural endowments.

Long-term Mitigation Scenarios (LTMS)

In 2007, the Long-term Mitigation Scenario (LTMS) project was published (See Chapter 2 of this book). This landmark project developed various scenarios using energy and macro-economic models to explore the consequences of various policy interventions at reducing GHG emissions (Winkler and Marquard, 2009). The LTMS proposed that the country peak its carbon emissions by 2025, plateau for a decade and then decline thereafter (DEAT 2007: 5). This process is known principally as 'peak, plateau and decline' (PPD), a common phrase in the country's climate change circles. According to the Department of Environmental Affairs (DEA), the PPD was defined as an upper limit of 428 million tonnes of carbon dioxide (Mt)/per annum in 2050 and a lower limit of 212 Mt/per annum CO_2. A scenario in the IRP update was that electricity-derived carbon emissions would peak around 275 Mt/pa CO_2 in 2022, plateau for 10 years, and start to decline around 2034. The decline scenario was either moderate, which would result in reaching 210 Mt/pa CO_2 in 2050, or an aggressive decline that would enable a low limit of 140 Mt/pa CO_2. However, it is estimated that the two new coal-fired power plants (Kusile and Medupi) and the return to service of three stations mothballed in the 1990s, are expected to have a combined coal consumption of over 50 Mt per year, adding more GHG emission in the near future (IEA, 2009). The approximate additional emission is 30 Mt of CO_2 annually (Rafey and Sovacool, 2011). The LTMS study was important because it showed clearly the requirements for the country to reduce its future carbon emissions. However, with two new coal plants, and more coal plants envisioned in the future, South Africa's energy policy can be characterised as somewhat contradictory: on the one hand, it promotes sustainable development whilst, on the other, it seems to be increasing its reliance on the extraction and consumption of polluting (and unsustainable) resources.

INSTITUTIONS

Mineral and Petroleum Resources Development Act (MPRDA)

Currently, the main legislative framework for mining and petroleum resources in South Africa is the MPRDA, which is under review. The most relevant section regarding sustainable development pertains to environmental authorisation and management. This section covers cross-cutting issues such as water, environment and licence to operate. According to the Department of Mineral Resources (DMR), it will designate itself as the Competent Authority for the implementation of the National Environmental Management Act (NEMA) on mining and prospecting sites, and the Minister of Environmental Affairs will be the appeal authority. The DMR has noted that a single piece of legislation implemented across departments will require resources and, more importantly, coordination, particularly on areas of monitoring and evaluation of compliance (DMR Annual Report 2013). Coordination is one of the critical issues in implementing policies and strategies in government. This requires inter- and intra-departmental coherent efforts where delineating ownership, responsibility and accountability are paramount. Furthermore, various amendments to the MPRDA do send signals of legislative uncertainty to the mines on the conditions they should operate in, as well as which policies are applicable. Thus, speedy resolution of these issues, which is a key factor in private sector decision-making, is critical.

Adding to this complexity is the need for nexus thinking, where interlinking issues need to be tackled holistically. According to the Mining Charter codes of good practice, environmental management is captured under sustainable development and growth. However, these standards are more reactive-based practices relative to compliance and monitoring and not enough towards pro-active measures. The DMR does have state-owned entities (SOEs) under its jurisdiction, such as Mintek and the Council for Geosciences, where excellent research and development are taking place. If a nexus approach were to interlink energy, water and waste as a means of ensuring sustainable development in the mining sector, more interdisciplinary research needs to take place in these areas. South Africa is not only a water scarce country with a carbon-intensive energy system, the legacy of mining is showing, with major signs of cumulative waste such as acid mine drainage. Solutions to these issues could enhance or complement a more efficient process as environmental externalities can no longer continue to be on the periphery.

Influence of the Business Lobby on Political Parties and Elections

South African legislation that currently governs elections is the Electoral Act of 1998 and the Public Funding of Represented Political Parties Act of 1997. The former requires parties to commit to: free electioneering, equal participation of women in political activities, allowing media access to electioneering activities, and full cooperation with the Independent Electoral Commission (IEC) (Lodge and Scheideggar, 2005). The latter determines the amount of money allocated to parties relative to the number of seats in both the national and provincial legislatures. The proportional amount accounts for 90 per cent of the funding and the remaining 10 per cent is on the basis of a threshold payment (Robinson and Brummer, 2006 and Lodge and Scheidegger, 2005). The main concern with regard to these two acts is the complete absence of matters pertaining to private funding. Lodge and Scheideggar (2005: 17) observed:

> *... there are no legal limits on how much parties can spend on electioneering or any requirements for disclosure of the sources or amounts of private donations ...*

Thus, rich benefactors can (intentionally or unintentionally) become indispensable sources of funding with hidden or visible motivations attached to such funding. According to Saffu (2002), in Africa, countries without disclosure laws outnumber those with such laws by a ratio of five to one. In the Southern African Development Community (SADC) region, the majority of member states (Lesotho, Mauritius, Malawi, Seychelles, South Africa, Tanzania, Zambia and Zimbabwe) do not have disclosure provisions for privately raised political party and campaign funds. Namibia is an exception – while it does not require full disclosure it obliges the disclosure of foreign donations. (Weissanbach, 2011) further pointed out that transitional states and young democracies tended to have fewer disclosure regulations. Of the 27 countries in Africa surveyed, 66 per cent did not have financial disclosure in place. This is indicative of apathetic accountability on this issue and it has serious implications, particularly for resource-rich countries. Collier (2007) stated that the natural resource trap is where bad governance in resource-rich countries leads to the stifling of the economy. Furthermore, there is a tendency for minimal regulation in these countries, undermining their capacity to gain full control of the sector. Combined with this is a huge propensity to rely on guaranteed sources of income. As such, the state is less held to account by the people.

In resource-rich countries, the implications of this abuse are increasing linkages between political sponsorship and the advocacy of self-interests, cronyism and party funding (Robinson and Brummer, 2006 and Dunnigan, *et al.* 2005). Besides direct contributions to the parties, it would seem that, in order to gain the sympathy of the ruling party, many companies could buy influence by transferring some portion of their assets to favoured groups in the context of Black Economic Empowerment (BEE).

One consequence of BEE is the emergence of a symbiotic relationship between those in power and those who are now empowered in important economic sectors. Robinson and Brummer (2006) indicated that with the allegiances between these two groups comes certainty, which is an enviable corporate asset. In South Africa, a study revealed that in the 1999 election period the political parties spent in the order of R300 million to R500 million on their election campaigns, but only R66 million came from public money (Robinson and Brummer, 2006). There is no legislation at present that forces political parties to declare their funding sources. The ANC secretary-general at the time, Kgalema Motlanthe (South African president from 2008 to 2009), revealed that the ANC was heavily reliant on donor funding. It was reported that the ANC receives the largest share of South African corporate funding amounting to 70 per cent (Lodge and Scheidegger, 2005). This report went further in identifying that the key donors, amongst others, were in the 'mining sector' (Lodge and Scheidegger, 2005: 18). Civil society is campaigning for transparency in the identification of funders, which could go a long way towards tackling the links between economic interests and political influence, to the extent that such influence can impact negatively on socio-economic policy, including matters to do with the environment. However, without disclosure laws, it is easier for businesses to field their interests through politicians by some form of influence peddling in exchange for guaranteed investment deals. Given the entrenched nature of mining and energy in the country's history, it should be appreciated that underlying interests in keeping the status quo are not easy challenges to overcome.

Integrated Resource Plan (IRP)

The Department of Energy (DoE) has updated its Integrated Resource Plan (IRP) with a decrease in demand from 454 terawatt-hours (TWh) to a range of 345 TWh–416 TWh by 2030 (DoE, 2013). Thus, a reduction of demand meant a decrease of 6.6 gigawatt (GW) of electricity capacity required. The

IRP update was very clear that modelling energy demand on the basis of an economy with high GDP growth and a significant shift away from energy-intensive industry would be difficult to accomplish (DoE, 2013: 24). What was more realistic was a lower GDP forecast with minor changes in industry structure. Thus, energy investment decisions had to take into account enough supply based on realistic demands, but not over supply that would result in stranded capacity (DoE, 2013: 24). Ultimately, the IRP update argued that the more aggressive the constraint was for carbon emission, which included carbon taxes, the greater the cost to the country's economic competiveness. This is an important narrative to take note of: the country does not foresee decoupling carbon emission from economic growth at a cost to overall development. Again, balancing key issues of competitiveness with climate change mitigation is a constant debate within the country. In other words, it would seem the sentiment is that appropriate economic growth for the country should not limit overall developmental goals and, while GHG mitigation is important, it may not be the most significant or decisive factor.

Financialisation

Financialisation is a term that is used by some to describe a combination of short-term capital inflows with a long-term outflow of capital through offshore listing of domestic corporations (Ashman, *et al.* 2011). The issue is important to this chapter because the MEC is a classic example of such finanicialisation. As South Africa's financial sector and system have matured (rated the strongest in Africa), so the financialisation of the MEC has become a dominant feature of the mining sector (Ashman, *et al.* 2011). The MEC developments have favoured short-term portfolio-based capital inflows rather than foreign direct investments, which are long-term in nature (Ashman, *et al.* 2014). The extent to which this has evolved is demonstrated by the size of stock portfolios the mining companies occupy within the Johannesburg Stock Exchange (JSE). The JSE market capitalisation is more than double the size of the actual economy, which is part of the global speculative commodity markets resulting in prices that are often not matched by production costs and demand (Isaacs, 2014). The MEC has historically favoured a more exclusive model – at first, the development and wealth creation of a white minority and, later, over the last 20 years, the new black elite as well as international capital. The political economy of the MEC itself, in turn, has a big influence on the structure and ethos of other parts of the economy as the interlinkages between elites are strengthened. An emphasis on a global commodity market renders the economy vulnerable to short-term volatile capital flows and less long-term labour absorbing economic growth. Essential for transformative change in the country is less focus on short-term shareholder value, addressing the dependency on the MEC, and identifying new areas of the economy that can be de-linked from the MEC.

South Africa faces huge problems that will take years to fix. These include a high current account deficit (6 per cent of GDP), low economic growth and continued labour friction, given infrastructure backlogs, an inadequate education system, and toxic labour relations. When there is macro-economic volatility, the rand exchange rate has to bear the brunt of this adjustment process (Isaacs, 2014). Furthermore, profits from the MEC need to be structured in a way that addresses long-term investment in the country and used appropriately to achieve development objectives, including the diversification of the economy.

INTERESTS

Mineral Energy Complex (MEC)

The MEC is a set of activities that locked in and co-evolved several industries together: intensive mining, mineral processing, the energy sector and associated industries. As a result, the MEC over time was central to the South African economy. Alongside this dominance is the integration of economic sectors across mining, manufacturing and finance. Therefore, the contribution of the MEC to the economy cannot be viewed in isolation, nor can it be measured in absolute terms. In a sense, South Africa suffers from a double 'lock-in' of a high carbon pathway dependency due to its dominant MEC, but also reliance on coal for its domestic energy derived from the MEC. The MEC also includes powerful and influential actors in the energy landscape. A system of accumulation of capital is manifested within the MEC, where consolidation of power in the hands of a few players has resulted in the creation of powerful and influential actors over energy and economic policy, as well as access to and distribution of investment (Takala, 2008). The MEC system had effectively incorporated different parts of the economic sectors through various forms of control, ownership and its relations with the state (Mcdonald, 2009). The dominance of the mining industry intrinsically gave way to a political economic system yielding to its requirements. This is reflected in energy policies that are geared towards promoting the interests of the MEC; these include macroeconomic stability that ensured suitable mineral exports, the creation of state-owned enterprises (such as Eskom), which are largely involved in minerals and energy, and a labour market that was used to ensure a steady supply of low-cost labour (Takala, 2008). Thus, the level at which the MEC is entrenched and fluid within the economic structure of South Africa poses a formidable challenge in finding a potential space for alternative sources of energy (other than fossil fuel).

Eskom

In order to consider sustainable energy development in the country, it is necessary to take into account the dominant relevance of Eskom. This company has a 95 per cent stake in the sector and is vertically integrated into the system. It is responsible for the generation, transmission and distribution of electricity. The national utility is one of the largest electricity utilities in the world. It was ranked thirteenth in the world by generation capacity in 2008 (EGI-SA, 2010). The majority of its customers are in mining, manufacturing and industry. Eskom has undergone a series of transformations over the years, particularly when it was converted into a tax-paying public company under the Companies Act, fully owned by the state (EGI-SA, 2010; Gentle, 2009; Gaunt, 2008). Eskom's direction changed from that of a public interest to that of a corporation.

Eskom had to ensure the shareholder's objectives were met, meaning that government and Eskom were now closely aligned (Gaunt, 2008). Conversely, the Municipal Electricity Undertakings (MEUs) are owned by the municipal councils, which, in their governance, are in turn accountable mostly to customers and voters in the local community, giving rise to a different set of objectives (Gaunt, 2008). Therefore, competing interests exist, particularly on issues surrounding increasing electricity costs. A mindset can then develop in which Eskom's main imperative is seen as ensuring profit while MEUs tend to address electricity price affordability. Given that Eskom is entrenched, and a significant monopoly in the energy system in the country, it will take some time to attain a different balance in the industry. In this regard, the Independent Systems Market Operator (ISMO) Bill has been proposed for more than a decade. It is a bill that would facilitate the introduction of Independent Power Purchasers (IPPs) to the system, providing competition in electricity generation, and result in the unbundling of the current monopoly. However, at present, the bill has almost been rejected in its totality because, among others, it would undermine the state's leverage in the energy industry and would cause too much disruption in the context of the current generation shortfalls. So the dominance of Eskom remains, although the recent introduction of IPPs may change the balance somewhat. The issue of decoupling state-owned energy supply to the mines also arises in this context. Furthermore, the issue of privatisation has been debated for over a decade, but the preference within the state is to retain the status quo rather than restructuring the entire electricity system. With the current challenge of load-shedding, the poor financial position of Eskom, lack of progress in new energy builds, and inconsistent leadership at Eskom, critical questions are starting to be asked about the proportion of IPPs in the system and funding models for Eskom, which may include a sale of part of the company's equity.

The ISMO bill is a good example of how competing interests continue to hamper energy reforms. As previously discussed, Eskom has had mutually reinforcing relationships between its large and intensive users, indicative of its historical ties to mining. Eskom depends on a few privately owned coal companies. Furthermore, the majority of its customers are represented by 31 large industrial consumers known as the Energy Intensive Users Group (EIUG), which accounts for 44 per cent of the electricity consumption (EGI-SA, 2010). The five biggest coal companies are Anglo-American, Xstrata, BHP Billiton, Sasol mining and Exxaro (Eberhard, 2011). Thus, the nucleus of economic growth and electricity infrastructure had started from Mpumalanga Province due to rich coal and mining activities, and then stretched to the main cities as the economy grew. As a result, the large industrial users have incurred minimal distribution costs and this, in essence, has created a two-tier system. Eskom has a first tier customer base of industrial users and a second tier service to the municipalities (Rustomjee, 2014). Thus, any major structural changes to Eskom would cause instability, price uncertainty and disruption of the mutually entrenched relationship with its existing users (Rustomjee, 2014), not to mention the possible economic impact it may have, as the main imperative at present is to ensure short-term electricity supply with future grid stability.

The current electricity crisis in the country is focusing energies on urgent short-term solutions, and sustainable development may not receive adequate attention and may be relegated to the long-term. Thus, managing between these time frames is one of the key issues that will require attention in defining energy projects. Without such a balance, the outcome may be to lock the country into business as usual with more fossil-fuel-derived resources such as coal. However, greener options towards more renewables are available and need to be pursued. Thus, despite well-informed reports such as the IRP update, or LTMS, with parallel uncertainties in future oil prices or global commodities demand, it should be appreciated that bold decisions are neither easy nor sequential. It is often repeated that vested interests among a few players, particularly associated with the MEC, are key to energy investment decisions. As long as these vested interests continue to dominate, and as long as they do not change their own paradigms, deviations from the norm will confront persistent barriers. In the recent period, a number of these large corporations have offered to contribute to electricity generation using gas (e.g. Sasol) or even the more innovative fuel cell technology which uses a significant level of platinum group metals (PGM). The latter, if it ever takes off in earnest, may open up possibilities for a new MEC relationship as it combines beneficiation with green technology (in that, the by-product of the fuel cell generation process is water).

Energy Security and Coal Exports

Another strong challenge for Eskom relates to the prospect of increased demand for low-grade coal. Traditionally, Eskom has had a reliable source of coal because it was of low quality and not suitable for export. South Africa depends on its export revenues and coal accounts for the third largest export after gold and platinum (Nkomo, 2009). Additionally, export revenues do not only fund the country's import expenses, they are also used for the development of new industrial sectors (NPC, 2011). This is a recurrence of the country's dependence on the MEC. The demand for RB3 low-grade coal has increased, particularly in countries such as India and China. The quantity of coal imported by India rose by eight per cent in March 2011 (Energy Global, 2011). Subsequently, Eskom is lobbying the government to recognise coal as a 'strategic national resource' ensuring that it takes priority relative to energy security (Creamer, 2011). Diversifying South Africa's use of resources away from coal and into renewable energy will take time, and so it will remain a critical asset. This reiterates the number of competing dynamics to be considered, both in the short- and the long-term.

Cosatu, Climate Change and the Green Economy

The Congress of South African Trade Unions (Cosatu), with approximately 1.8 million members (StatsSA, 2005), is the largest trade union federation. Until recently, the single biggest affiliate was the National Union of Mineworkers (NUM) with 300,000 members. The largest union affiliate is now the National Union of Metalworkers South Africa (Numsa), which, at the time of publication, was fighting its expulsion from the federation. Cosatu has a structured alliance with the ruling party, namely the African National Congress (ANC) and the South African Communist Party (SACP), which together form the Tripartite Alliance (Piper and Matisonn, 2009; Webster, 1998; Mckinley, 2001). There are major policy dialogues that occur within this alliance and it is an important source of information for the current ruling party, the ANC (Mohanty, *et al.* 2011; Webster and Buhlungu, 2010; Buhlungu, 2008).

According to Cherry (2006), Cosatu has a multifaceted interaction with, and influence on, government by means of the following:

- Advocacy (for example social security for the poor, job creation and privatisation)
- Negotiations in the National Economic Development and Labour Council (Nedlac)
- Working through political processes of the Tripartite Alliance
- Mass mobilisation (such as action through collective strikes)

These structures of communication cannot be understated because they can have significant implications when it comes to implementing policies that may or may not be welcomed by Cosatu. There is a tendency for the majority of the people to support mass action as a complementary strategy for putting pressure upon elected politicians if the government fails to deliver promised benefits, most notably demonstrated in 'service delivery protests' (Mohanty, *et al.* 2011 and Buhlungu, *et al.* 2008). Mass and repeated strike action over prolonged periods of time can cripple the economy. Thus, Cosatu is central to the ongoing dynamics within government policies and strategies, especially at the grass roots level, as well as government, industrial sectors, and civil society.

The Mixture of Business and Politics

The most significant development in this sector has been BEE deals which have resulted in black-owned coal companies controlling more than 30 per cent of South African coal production. The largest BEE deal of 2003 illustrates the lack of breadth in the empowerment process. The deal was between ARMGold and Avmin for R10.6 billion. Analysis revealed that at least 72 per cent of the total deal value benefited six narrow-based black consortia (some of which have politically well-connected beneficiaries) (Dunnigan, 2005). The emergence of large black operators in an economy that had been solely dominated by whites is not a backward step as such. However, the transfer of wealth from one elite group to another – without a structural change in the path dependency – mostly perpetuates old patterns of social inequality. The system of wealth accumulation and the continued privilege of elites need to be confronted in the broader process of transformation. This skewed distribution of wealth has also resulted in the creation of powerful actors that are influential in determining the energy policies of the country. Moreover, this lopsided distribution of wealth may in fact keep the division of the 'two world economy'. The two world economy refers to one end of the spectrum where South Africa has high living standards and access to infrastructure (which includes the basics such as electricity, water and sanitation) that are comparable to those in more developed countries (Mcdonald, 2009 and Winkler and Marquard, 2009). However, at the other end is the poor population who lack access to basic living standards (housing, health, education and energy). This disproportionate situation is primarily due to the legacy of apartheid, which has embedded many socio-economic challenges and issues.

WAY FORWARD

As a way of understanding this multitude of issues, a framework by Voss and Kemp, 2005, is discussed. They proposed a reflexive governance approach whereby formulating policies should be an exercise of flexibility and adaptability. Since sustainable energy development is a complex issue, knowledge from various sources is required. Knowledge production cannot only rely on scientific sources; it also needs to take into account the needs of societal actors (Voss and Kemp, 2005). There is a need to create 'integrated knowledge production' that enables information, not only from within the boundaries of science and policy, but transcends these towards society in general (Voss and Kemp, 2005: 10). In terms of uncertainty, future energy development cannot be predicted with precision, nor can it be predicted for its systemic effects. Therefore, a level of 'adaptivity' and flexibility is needed relative to the institutions and technological systems that govern implementation (Voss and Kemp, 2005: 11). An important factor here is to *capacitate* the system to respond to changes adequately and proactively. Moreover, a continuous system of *monitoring* and *evaluation* needs to be in place once policy implementation starts.

Energy investments have a long lifespan and changing trajectories is not easy as prior technologies are path dependent or lock-in for a certain timeframe. Specific patterns are realised and stabilise within social values, institutions and societal systems. An example here is coal-derived electricity. Most people need a consistent and reliable electricity supply, and whether the source is carbon-intensive is usually a secondary matter. Therefore, an energy technology choice needs to take into consideration the long-term systemic effects. South Africa continues to be locked into a carbon-intensive energy system, and some may argue that business as usual needs to continue because coal is abundant and cheap. However, South Africa needs to ask itself whether current efforts to gradually extricate itself from this carbon-intensive system are adequate, and how stakeholders within the energy and sustainable development community can work together towards optimal solutions.

Strategy implementation requires *meta-coordination*, not only within and among government departments, but also with various stakeholders. Major factors to consider for implementation are: 1) the project must have a defined champion with clear mandates; 2) there should be sufficient institutional capacity where projects are aligned to mainstream and existing initiatives at scale; 3) create a conducive and supportive context for policy, regulation and planning, and ensure ownership of the integrated process; 4) ensure that projects are aligned to national priorities which should assist in mainstreaming and alignment with broader mandates or enhance existing mandates and, lastly; 5) there must be sufficient technical capacity to enable the project to be sustainable.

Among the measures required to reduce the country's dependency on the MEC would be to take up in earnest the opportunities opened up by the knowledge-based economy. This will ensure that the country is more resilient to volatile commodity markets and is less reliant on primary resources. The experience from the mining industry can stand the country in good stead, particularly with regard to the knowledge base that includes production or process engineering, as well as research and development. An important consideration is to focus not only on the continuous output of mining resources and seeing the minerals as an end in themselves, rather, the knowledge underlying the processes of obtaining these minerals should be utilised to diversify or enhance other sectors of the economy. Examples could include technological innovation in mechanisation process, machineries, parts, tools and equipment – all of which could spill over into important sectors such as transportation, information technology, construction and manufacturing. Good economic policy and planning that have clear growth and development objectives should dictate how mining investments flow and how revenues or income generated from the mining sector are managed for the benefit of society as a whole.

CONCLUSION

There is no doubt that mining in the country will continue to be a dominant sector in the years to come. However, the sector is undergoing major challenges that cut across significant issues that are at the heart of the country's developmental goals that include labour, economic growth and unemployment. The role of the MEC is indicative of the entrenched regime that the mining sector plays within the economy. The MEC has broadened itself into a system of wealth accumulation and has integrated large sectors that include mining, transport, manufacturing and finance. As was discussed, the EIUG is also one of the biggest companies in the mining sector, as well as the JSE. Due to a system of wealth accumulation, there is a handful of players in this sector and thereby also a few powerful actors that are influential in determining key policies and strategies. Thus, the argument in this chapter is that vested interests within the MEC are key in understanding the implementation of energy policies in the country.

The issue of ideological mismatch between a developmental state and a more neo-liberal economy is a fundamental obstacle. The influence and the role of trade unions cannot be overstated: major policy dialogues take place within the Tripartite Alliance. Unbundling Eskom has been stalled several times and is part of a larger privatisation debate that encompasses the nationalisation of mines and selling off some of Eskom's assets. The ideological mismatch has to be tackled because it is a root cause for policy incoherencies and implementation inertia.

The subject of financialisation is indicative of a need for major reform in the structure of the economy. Investments made by the mining sector that favour profits flowing out of the country instead of long-term financial investments are compounding developmental challenges. These include uneven wealth distribution mostly skewed towards elitism and wealth concentration, thus inequality persists. The JSE total market capitalisation is twice the size of the actual economy. The emphasis on a commodity market renders the economy vulnerable to short-term volatile capital flows, and not necessarily long-term labour absorbing economic growth. The current energy crisis provides an opportune moment to revisit the country's mining economic policy.

References

African National Congress (ANC). 2012. 'State Intervention in the Minerals Sector (SIMS)'.

Ashman, S., Fine, B. and Newman, S. 2011. 'The Crisis in South Africa: Neoliberalism, Financialization, and Uneven and Combined Development', *Socialist Register*, 47: 174–195.

Besada, H. 2007. 'Fragile Stability Post-Apartheid South Africa'. Working paper no. 27, Centre for International Governance Innovation (CIGI).

Bronkhorst, Q. 2014. 'These are the biggest listed companies in South Africa'. Available at: http://businesstech.co.za/news/general/59253/these-are-the-biggest-listed-companies-in-sa/ [Accessed 1 December 2014].

Buhlungu, S. 2008. 'Gaining influence but losing power? Cosatu members and the democratic transformation of South Africa', *Social Movement Studies*, 7(1): 31–42.

Buhlungu, S., Southall, R., Webster, E. 2008. 'Cosatu and the democratic transformation of South Africa' in Buhlungu, S. (ed.) *Trade Unions and Democracy, Cosatu Workers Political Attitudes in South Africa*. Cape Town: HSRC Press.

Chadbourne, 2014. 'Renewable energy near mines'. Available at: ttp://www.chadbourne.com/Renewable_Energy_Near_Mines_projectfinance/ [Accessed 1 June 2014].

Collier, P. 2007. *The Bottom Billion, Why the Poorest Countries are Failing and What Can be Done About It*. Oxford, UK: Oxford University Press.

Creamer, T. 2011. 'Eskom lobbying for a new coal dispensation to shore up domestic supply', *Engineering News*. Available at: http://www.engineeringnews.co.za/article/eskom-lobbying-for-a-new-coal-dispensation-to-shore-up-domestic-supply-2011-01-26 [Accesssed 1 March 2011].

Davidson, O. 2006. 'Energy Policy' in Winkler, H. (ed.), *Energy Policies for Sustainable Development in South Africa. Options for the Future*. Energy Research Centre (ERC),

University of Cape Town, South Africa.

Department of Energy (DoE). 2010. *Strategic Plan 2010/11–2012/13*. Pretoria, South Africa.

Department of Energy (DoE). 2012. 'Draft of the Second National Energy Efficiency Strategy of the Republic of South Africa', Pretoria, South Africa.

Department of Energy (DoE). 2013. *Integrated Resource Plan for Electricity (IRP)*. 21 November 2013 update.

Department of Environmental Affairs and Tourism (DEAT). 2007. 'Long-term Mitigation Scenarios (LTMS)', Technical Summary, Pretoria, Scenario-building Team (SBT), Pretoria.

Dunnigan R., Fazaeli, K., Spies, J. 2005. 'Black economic empowerment – difficulties and opportunities in making right the wrongs of the past'. Available at: http://faculty-course.insead.edu/dutt/emdc/projects/Sep-Oct05/Group_D.pdf [Accessed 1 March 2012].

Eberhard, A. 2011. 'The future of South African coal: market, investment, and policy challenges', *Working Paper 100, Program on Energy and Sustainable Development (PESD)*. Stanford, California, USA: Stanford University.

Electricity Governance Initiative – South Africa, (EGI-SA). 2010. Lowy Institute for International Policy. Cape Town: IDASA.

Fine, B. and Rustomjee, Z. 1996. *The Political Economy of South Africa: From Minerals Energy Complex to Industrialization*. London: C. Hurst.

Gaunt, C. T. 2008. 'Electricity distribution industry restructuring in South Africa: a case study', *Energy Policy*, 36: 3448: 3459.

Geels, F. W. and Schot, J. 2010. 'The Dynamics of Transitions: A Socio-Technical Perspective', in Grin, J., Rotmans and Schot, J. (eds.) *Transition to Sustainable Development, New Directions in the Study of Long-term Transformative Change*. New York: Routledge.

Gentle, L. 2009. 'Escom to Eskom: from racial Keynesian capitalism to neo-liberalism (1910–1994)' in McDonald, D.A. (ed.) *Electric Capitalism, Recolonising Africa on the Power Grid*. UK: Earthscan, Dunstan House.

Giddens, A. 1984. *Constitution of Society*. Berkley and Los Angeles, CA, USA: University of California Press.

Government Communication Information System (GCIS). 2012. 'Mineral resources', *Pocket Guide to South Africa 2011/2012*.

Harrison, K. and Sundstrom, L. M. 2007. 'Introduction: The comparative politics of climate change', *Global Environmental Politics*, 7 (4): 1–18.

Heun, M. K., Nieker, J. L., Swilling, M., Meyer, A. J., Brent, A., Fluri, T. P. 2010. 'Learnable lesson on sustainability from the provision of electricity in South Africa, Proceedings of the ASME 2010', *4th International Conference on Energy Sustainability*. Phoenix, Arizona, USA, 17–22 May 2010.

Hodgson, G. 2006. 'What are institutions?' *Journal of Economic Issues*, XL: 1–25.

International Energy Agency (IEA). 2009. 'Key world energy statistics, 2008', International Energy Agency, Paris, pp. 14–15.

Issacs, G. 2014. 'The Financialization of the South African Economy and the Havoc it Wreaks'. Available at: http://www.sacsis.org.za/site/article/1959 [Accessed 1 June 2014].

Kotze, C. 2014. 'JSE resource listings plunge 50% in last 20 years but total market cap now a massive R2.82 trillion'. Available at: http://www.miningweekly.com/article/number-of-jse-listed-miners-decrease-while-market-capitalisation-in-resource-sector-increases-2014-08-08 [Accessed 1 December 2014].

Lazenby, H. 2014. 'Six cronimet projects progress to next South African IPP round'. Available at: http://www.miningweekly.com/article/six-cronimet-projects-progress-to-next-south-african-ipp-round-2014-03-10 [Accessed 1 December 2014].

Letsoalo, M., Mataboge, M., Huner, Q. 2014. 'Cosatu split, how the ANC will suffer'. Available at: http://mg.co.za/article/2014-11-13-cosatu-split-how-the-anc-will-suffer [Accessed 1 December 2014].

Lodge, T. and Scheidegger, U. 2006. 'South Africa country report based on research and dialogue with political parties', International Institute for Democracy and Electoral Assistance (IDEA).

McDonald, D. A. 2009. *Electric Capitalism, Recolonising Africa on the Power Grid*. UK: Earthscan, Dunstan House.

McKinley, D. T. 2001. 'Democracy, power and patronage: debate and opposition within the African National Congress and the Tripartite Alliance since 1994', *Democratization*, 8 (1): 183–206.

Mohanty, R., Thompson, L., Coelho, V. S. 2011. 'Mobilising the state? Social mobilization and state interaction in India, Brazil and South Africa', *Institute of Development Studies (IDS) Working Paper*, 359.

Nkomo, J. 2009. 'Energy and economic development' in Winkler, H. (ed.) *Energy Policies for Sustainable Development in South Africa. Options for the Future*. Energy Research Centre (ERC), University of Cape Town, South Africa.

Paton, C. and Marrian, N. 2014. 'Cabinet leaning to break up Eskom', *Business Day*. Available at: http://www.bdlive.co.za/business/energy/2014/08/05/cabinet-leaning-to-break-up-of-eskom [Accessed 1 December 2014].

Pegels, A. 2010. 'Renewable energy in South Africa: Potentials, barriers and options for support', *Energy Policy*, 38: 4945–4954.

Piper, L. and Matisonn, H. 2009. 'Democracy by accident: The rise of Zuma and the renaissance of the tripartite alliance', *Representation*, 42 (2): 143–157.

Rafey, W. and Sovacool, B. K. 2011. 'Competing discourses of energy development: The implications of the Medupi coal-fired power plant in South Africa', *Global Environment Change*, 21 (3): 1141–1151.

Raubenheimer, S. 2011. *Facing Climate Change: Building South Africa's Strategy*, IDASA. Cape Town: Unity.

Robinson, V. and Brümmer, S. 2006. 'SA democracy incorporated. Corporate fronts and political party funding', Institute of Security Studies (ISS) paper 129.

Rustomjee, Z. 2014. 'Vested interests have hijacked energy policy', *Business Day*, Opinion and Analysis, 19 December.

Saffu, Y. 2002. 'Where money meets politics: Exploring policy options to regulate the influence of private funders on South Africa politics', Paper prepared for the ISS seminar, 31 October – 1 November 2002, Leriba Lodge, Centurion, South Africa.

Sasol. 2010. 'Annual review and summarized financial information, focused and energized'.

Singh, S., 2014. 'Effects of the mining strikes on the South African economy'. Available at: http://www.miningreview.com/effects-of-the-mining-strikes-on-the-south-african-economy/ [Accessed 20 December 2014].

Stedman, S. J, 1994. *South Africa, the Political Economy of Transformation*. Boulder, Colorado, USA: Lynne Rienner Publishers.

Takala, L. 2008. 'The role of industrial policy and the minerals-energy complex in the decline of South African textiles and clothing'. The 11th Annual Global Conference of The Competitiveness Institute (TCI), Cape Town, South Africa, 27–31 October 2008.

United Nations Statistics Division, (UNSD). 2008. 'Millennium Development Goals indicators: Carbon dioxide emissions (CO_2)'.

Voss, J. P. and Kemp, R. 2005. 'Reflexive governance for Sustainable development – incorporating feedback in social problem solving'. Paper for International Conference of the European Society for Ecological Economics (ESEE) Lisbon, Portugal, 14–17 June 2005.

Webster, E. C. 1998. 'The politics of economic reform: Trade unions and democratization in South Africa', *Journal of Contemporary African Studies*, 16(1): 39–64.

Webster, W. and Buhlungu, S. 2010. 'Between marginalization and revitalisation? The state of trade unionism in South Africa', *Review of African Political Economy*, 31 (100): 229–245.

Weissenbach, K. 2011. 'Party finance and party competition in dominant multi-party systems: the case of South Africa', paper presented at the IPSA-ECPR Joint Conference: *What happened to North-South?* 16–19 February. Sao Paulo, Brazil: University of Sao Paulo.

Winkler, H. 2005. 'Renewable energy policy in South Africa: policy options for renewable electricity', *Energy Policy*, 33: 27–38.

Winkler, H. 2009. *Cleaner Energy Cooler Climate. Developing Sustainable Energy Solutions for South Africa*. Cape Town, South Africa: HSRC Press.

Winkler, H. and Marquard, A. 2009. 'Changing development paths: From an energy-intensive to low-carbon economy in South Africa', *Climate and Development*, 1: 47–65.

World Bank. 2010. Available at: http://data.worldbank.org/indicator/EN.ATM.CO2E.PC [Accessed 1 May 2015].

World Energy Council. 2010. *Survey of Energy Resources*, 22nd edition. London, United Kingdom.

CHAPTER 4

Lost in Procurement: An Assessment of the Development Impact of the Renewable Energy Procurement Programme

Fumani Mthembi

> *... but is it [development] that is being understood here or is its identity being evaded in reducing it to a subset of practicable measurements?*[29]

ABSTRACT

This chapter analyses the socio-economic development impact of the Renewable Energy Independent Power Producers Procurement Programme (REIPPP) in communities where renewable energy farms are located. A number of limitations were observed and these limitations result from a range of factors such as incomplete development targets, the reluctance of independent power producers to make investments in socio-economic development, an absence of a shared understanding of development and, most critically, a weak monitoring system. Therefore, this chapter argues that the

29. Adapted from Luke, T. W. 1995. 'On Environmentality: Geo-Power and Eco-Knowledge in the Discourses of Contemporary Environmentalism', *Cultural Critique, No. 31, The Politics of Systems and Environments, Part 3*, pp. 57–81 'but is it the environment that is being understood here or is its identity being evaded in reducing it to a subset of practicable measurements?'

design of the on-grid procurement system, which explicitly and laudably sets out development targets for independent power producers, is undermined by its conflation of compliance with development impact. In other words, the failure to tease out the meaning and full potential of development through this sector is producing results that fall short of and, in some instances, completely subvert the transformation intent behind the state's procurement of on-grid renewable energy. The chapter identifies 10 key themes to demonstrate the missed opportunities with respect to development since the sector's inception in 2011. Based on this, solutions are proposed that incorporate an expanded interpretation of development into the existing monitoring system to ensure that the state is measuring the right factors and independent power producers are incentivised to invest in and account for impactful rather than check-box development.

INTRODUCTION

Development can be defined in multiple ways. Cowen and Shenton argue that 'one of the confusions, common through development literature is between development as an immanent and unintentional process … and development as intentional activity'.[30] In truth, the alternatives presented by Cowen and Shenton represent the extreme ends of a single spectrum. In other words, this characterisation is almost akin to free market versus state-controlled economics, with mixed economy variations lying somewhere in between. What this chapter is assessing – the Renewable Energy Independent Power Producer Procurement Programme (REIPPP) – resides in the 'mixed economy' zone as – although defined by the state – it is an initiative aimed at liberalising South Africa's energy sector by including private companies in power generation. In other words, our concern is intentional development.

What this chapter seeks to demonstrate is that, in some cases, the Department of Energy has not succeeded in matching development intent with development practice and that, while in the short run the negative impacts of these shortcomings are not obvious, the long-run social risks stemming from this dissonance are substantial. Indeed, the title phrase of this chapter, 'Lost in Procurement', is a pun derived from the expression 'lost in translation', which indicates an instance in language translation that does not allow for the full meaning of one language to be captured in another. Similarly, it is argued in this chapter that the full meaning of development is lost or compromised when it is

30. Cowen M. and Shenton, R. 1998. *Doctrines of Development*. London: Routledge

translated into the current language of compliance, which is expressed through the REIPPP's Economic Development framework.

More fundamentally, however, it is important to appreciate that the notion of development is value-laden. Its articulation is a function of how society perceives itself and its related, collective aspirations.[31] To determine the meaning of development, this chapter anchors itself in South Africa's vision for itself as expressed in the National Development Plan: '... by 2030, we seek to eliminate poverty and reduce inequality. We seek a country wherein all citizens have the capabilities to grasp ever broadening opportunities available.'[32] Therefore, this chapter's analysis of the DoE's economic development framework, as set out for the on-grid renewable energy sector, seeks to determine whether each target and, crucially, the observed practices linked to the implementation of each target, make our intent, our national vision for development, more or less attainable.

What is patently clear is that 'you cannot manage what you don't measure' and, therefore, it is not compliance as the act of measuring that this chapter argues against. Instead, it is the assumption that the current compliance framework as expressed through Economic Development (ED) obligations is a sufficient means for the attainment of development. The chapter therefore assesses ED in terms of poverty reduction, the expansion of capabilities and long-term sustainability of socio-economic investments. In doing this, the chapter seeks to unpack and propose solutions to how to best deliver value for money through the current economic development framework. It is necessary to state upfront that the current ED framework is seen as a sound foundation or starting point. What is thus suggested is augmentation rather than a complete overhaul of what is already in place. This point is supported by other researchers who, having assessed South Africa's procurement legislation, conclude that what is necessary is improvement upon what exists rather than the creation of something entirely new.[33]

What we seek to critically discuss, then, is state procurement in the context of the Renewable Energy (RE) sector. Renewable energy, being newly liberalised in South Africa,[34] presents an opportunity to achieve greater development impact through the design, implementation and monitoring of the sector's procurement system. This system, REIPPP, which began in 2011, is

31. Chambers, R. 1997. *Whose Reality Counts? Putting the First Last*. London: ITDG.
32. National Planning Commission. 2011. *National Development Plan*.
33. Turley, L. and Perera, O. 2014. 'Implementing Sustainable Public Procurement in South Africa: Where to Start?' IISD Report.
34. South Africa's first Independent Power Producers in the Renewable Energy Sector were decided in November 2011 upon the awarding of preferred bidder status to 28 bidders. Of these, 27 went on to be granted Power Purchase Agreements, licences for energy production in December of 2012.

a competitive bidding process through which the state, as the sole buyer of energy for the national electricity grid, assesses the proposals of Independent Power Producers (IPPs). The state's assessment is based on two factors: price, which is linked to finance and technology, as well as economic development, which is comprised of seven elements (which will be discussed in the background section of the chapter). For the purposes of this chapter, it is economic development (ED) that is placed in the spotlight in order to make evident the nature and urgency of the misalignment between the intentions of the programme and the practices that pervade its implementation. Each element of economic development is dealt with in detail, barring local content. This issue requires dedicated separate attention given its interplay with local manufacturing and thus, the Industrial Policy Action Plan (IPAP 2).

RESEARCH AND ANALYSIS APPROACH

Because the sector is new, many assessments have tended towards descriptions of its machinations and its expected outcomes based on bid submissions. For example, Eberhard, *et al.* published a paper in 2014 titled 'South Africa's Renewable Energy IPP Procurement Programme: Success Factors and Lessons', which provides a comprehensive history of REIPPP and includes statistical data on the sector's composition and trends thus far. The economic development impact of the sector, which is linked to a 20-year outlook (pegged to the length of Power Purchase Agreement (PPA)), has not been dealt with in much detail owing to an absence of data. It is true that there is not sufficient data at this point to comment on whether the right social investments were made towards community development, for example.

However, what are immediately observable are sector practices in relation to economic development. It is these practices that are already providing a rich set of clues into how various stakeholders perceive development. They also reveal weaknesses in the DoE's monitoring mechanisms and, critically, point towards the generation of social risks that cannot be captured or punished using the current measures. As is known, it is the protracted under-appreciation of social risk that results in violent and expensive forms of social upheaval. Certainly, cases such as Marikana[35] reveal that a dogmatic marriage to suboptimal technocracies will, in the long run, result in total breakdown. Therefore, in proposing changes to the current ED framework, we propose openness in principle to constant adjustment in the ways in which ED is monitored to

35. For a summary of initial Marikana Commission outcomes, see: http://mg.co.za/article/2014-09-16-key-lessons-for-lonmin-as-marikana-commission-wraps-up

ensure that social risks don't lead to costly eruptions.

But how exactly do we come to grips with sector practices in a context where competing private actors are not compelled to share how they execute their plans? This speaks to the data collection challenge at the heart of this chapter. Data was collected by researching communities based in Limpopo, North-West, Free State, Northern Cape and Western Cape and the impact of these programmes understood.

The chapter opens with a detailed description of economic development, identifying each element and how the state measures compliance. It follows with an assessment of 10 key hurdles to development. These hurdles identify different types of design and implementation failures that undermine the ultimate goal of development. Associated with each hurdle is a proposed solution as well as further research questions that must be pursued to strengthen the scientific understanding of how widespread certain forms of malpractice are, as well as the potential veracity of the solutions that the chapter points to.

BACKGROUND: ECONOMIC DEVELOPMENT IN THE RENEWABLE ENERGY SECTOR

To start with, it is critical to understand the timetable of renewable energy power plants to gain perspective on when and how the social risks identified in this chapter unfold. There have been four bidding windows each year since the inception of REIPPP in 2011. The success of a bid is typically announced three to six months after the bid submission, after which another six-month period is granted for Independent Power Producers to reach financial close. At this point, they are awarded Power Purchase Agreements (PPA) by the state. Construction typically begins three months after the award of a PPA and lasts, on average, for 18 months, depending on the size of the power plant and the speed at which it reaches Commercial Operation Date (COD), which is a technical milestone determined by the state. In all then, it can take up to three years for a power plant to become fully operational. Therefore, of the first three bidding rounds that have been awarded, only Round 1 (28 awarded) and Round 2 (19 awarded) projects have been granted power purchase agreements. Furthermore, while Round 1 projects have largely reached the operations phase, Round 2 projects are mostly in the final phases of construction. The 17 Round 3 projects that were successful have not yet signed their PPAs with the government.

The key implication of this timetable is that many of the development risks that are identified fall into the realm of missed opportunities. For example, the failure to train workers during construction is not likely to result in a workers' protest because workers do not understand this to be an entitlement, but serves as a missed development opportunity given the unchanged skills profile of the labour force. The longer-term issues, such as the inability to generate a class of Black industrialists or the absence of communities from the planning process, are issues that will manifest over the course of the 20-year life cycles of the projects. Thus, this chapter is about understanding the foundational deficiencies that may compromise development in the long run.

Economic development in the renewable energy sector is 30 points out of a possible 100, which means that, when bidding for a licence to supply electricity to the national grid, each bidder should focus 30 per cent of its energies on matters of ED.

Economic development is disaggregated into seven areas, known as pillars:

- job creation;
- local content;
- preferential procurement;
- top management;
- ownership;
- enterprise development, and
- socio-economic development.

Upon granting a licence,[36] the IPP is expected to fulfil the ED obligations it committed to in its initial bid and is subsequently monitored on a quarterly basis by the DoE.

Put differently, each element of economic development constitutes a performance commitment to the state. The DoE determines minimum standards known as 'thresholds', and measures the quality of a bid on the IPP's ability to not only meet but also exceed the threshold requirements. The table below is a summary of how each economic development element is measured by the DoE. It does not include actual thresholds or targets as these change with every bidding round. It is thus intended to provide an understanding of what is measured and how it is measured. It is worth stating that this chapter does not interrogate whether the DoE's current thresholds are set at a level that

36. The term 'licence' is used loosely to mean the signing of multiple agreements between the DoE and IPPs, which grant IPPs permission to construct and operate their planned power plants and bind the state to the purchase of the resultant energy.

is consistent with national imperatives. Rather, we challenge the issues surrounding these targets to indicate that, regardless of how they have been set, the prospects of using them to approach true development are compromised under the current conditions.

ED Summary Table

The table below provides a summary of each ED element that is measured by the DoE, indicating the calculation method as well as the evidence that the DoE requires to monitor the performance of IPPs.

ED Element	Measure	Phase of Measurement	Evidence of Activity
1. Job Creation		Construction and 20-year operation	
South African citizens	Per cent of employees who are South African citizens relative to total employees measured in terms of time worked.		- Letter of appointment - ID - Time sheet - Salary advice
Black citizens	Per cent of Black citizens relative to total employees measured in terms of time worked.		
Skilled Black citizens	Per cent of skilled Black citizens relative to total skilled employees measured in terms of time worked.		
Local community	Per cent of employees from within 50km of project site measured in terms of time worked.		
Jobs per megawatt	Total time worked by total workforce divided by the total megawatts of a power plant.		

2. Local Content	Per cent of construction costs spent on South African goods and services.	Construction	- Invoice - Proof of payment - Local content declaration
3. Preferential Procurement	Total expenditure on South African goods and services.		- Invoice - Proof of payment - BBBEE certificate of supplier
BBBEE	Per cent of procurement of goods or services from enterprises that do not qualify as QSE or EME, measured in terms of the total Rand value relative to total procurement spend.		
QSE/EME	Sum of expenditure on QSEs and EMEs; Per cent of procurement from Qualifying Small Enterprises (enterprises that generate less than R5 million in annual revenue).		
Women-owned vendor	Per cent of procurement of goods or services from enterprises that are owned by women (51 per cent or more), measured in terms of the total Rand value relative to total procurement spend.		
4. Top Management		Construction & 20-year Operation period	
Black top management	Per cent of Black South Africans who constitute part of the top management team.		- Letter of appointment - ID - Time sheet - Salary advice

5. Ownership			
Black ownership in the seller	Per cent of Black ownership in the Independent Power Producer. Benefit realised through IPP dividends, if and when declared.	Construction and 20-year operation period	- IDs of black owners - Shareholders' agreements - Shareholders' certificates
Local community ownership in the seller	Per cent of local community ownership in the Independent Power Producer, held through a community trust for the benefit of the community within a 50km radius of the power plant. Benefit realised through IPP dividends, if and when declared.	Construction and 20-year operation period	- Community trust deed
Black ownership in the construction contractor	Per cent of Black ownership in the company appointed to construct the power plant. Benefit realised through IPP dividends, if and when declared.	Construction	- IDs of black owners - Shareholders agreements
Black ownership in the operations contractor	Per cent of Black ownership in the company appointed to operate the power plant over its 20-year life. Benefit realised through IPP dividends, if and when declared.	Operations	- IDs of black owners - Shareholders agreements
6. Enterprise Development	Per cent of annual revenue spent on the development of enterprises. The same expenditure is recognised differently depending on where recipients of the support are from, with the highest reward being for investing in the local	Operations	- Proof of funds disbursement - Identifying documents for fund recipients - Evidence of purpose of funds disbursement

	community (defined as being within a 50km radius of the power plant) and the lowest for investments outside South Africa.		
7. Socio-economic Development	Per cent of annual revenue spent on the socio-economic development investments that enhance the economic participation of previously excluded groups. The same expenditure is recognised differently depending on where recipients of the support are from, with the highest reward being for investing in the local community (defined as being within a 50km radius of the power plant) and the lowest for investments outside South Africa.	Operations	- Proof of funds disbursement - Identifying documents for fund recipients - Evidence of purpose of funds disbursement

Note: The above table is derived from the Renewable Energy Independent Power Producer Request for Proposal, which has been in existence since 2011, with minor adaptations in each subsequent bidding round.

The narrative and table preceding thus provide a summary of economic development in the renewable energy sector. Each of the seven categories that the state identifies is measured to determine compliance with the state's rules in the bidding phase and, thereafter, for every quarter starting from the power plant's construction to the conclusion of its 20-year operations phase. Given that the sector is new in South Africa, it is only a handful of power plants that were awarded licences in 2012 that are now concluding construction and entering the operations phase. Thus, while this chapter identifies oversights and missed opportunities thus far, it is important to recognise that there is still a great opportunity to adapt the Procurement Programme given that we are still at the sector's infancy.

The next section references the ED framework described above to provide a detailed account of how flaws in the current design, interpretation and monitoring of REIPPP have led to practices that compromise development.

HURDLES TO DEVELOPMENT: UNDERSTANDING THE DIFFERENCE BETWEEN COMPLIANCE AND DEVELOPMENT

The previous section has detailed the ED compliance framework as it is currently defined and measured. This section demonstrates that this framing is narrow, often resulting in the contradictory outcome of full ED compliance and partial, if not zero, actual development.

It is therefore proposed that development in REIPPP should be understood in terms of seven 'equations' which link the current ED compliance elements with development outcomes that can and should be measured.

The proposed seven Development Equations are as follows:

- Job Creation ***PLUS*** Skills Development
- Procurement ***PLUS*** Supplier Development
- Ownership ***PLUS*** Operational Involvement
- Management ***PLUS*** Key Roles for Black/South African Managers
- Socio-economic and Enterprise Development ***PLUS*** Community Participation
- Development Spend ***PLUS*** Impact Measurement
- Measuring ***PLUS*** Management

What is demonstrated in the coming subsections, through 10 topics that have been titled 'The 10 Hurdles', are the current compliance practices that are at odds with development. This discussion makes evident the conceptual gaps in the design of REIPPP and demonstrates how these gaps have resulted in practices that in fact undermine national development aims. By doing this, the intention is to make clear the case for 'The Seven Equations of Development'. The end of this section will suggest a practical way forward, which clearly shows how to incorporate the proposed 'Seven Equations of Development' into the existing framework to ensure that development is both measured and managed in the sector.

Hurdle 1: Exclusion of Development from the Framing of Risk

The notion of risk is central to the RE Procurement Programme. Technology and finance are vetted on multiple levels, starting with the owners, to the banks that fund the projects, to the development financiers that fund Community Trusts and, ultimately, by the Department of Energy. Much of this is in aid of balance sheet protection, including the most important balance sheet of all, the national fiscus. And it is only right that this should happen because this programme entails the expenditure of billions of rand that should be carefully guarded.

The problem, however, is that the risks of bad development are seldom considered in the planning and vetting of RE projects. This has remained the case since the start of the programme because the outcomes of bad development are generally expressed at the level of the communities that host RE power plants. By definition then, the effects of bad development are not fully known by Johannesburg and Cape Town-based managers and the affected, often-remote communities are unaware of their rights or potential for recourse. Below is a consideration of bad development at community level, which provides an indication of the major risks that currently simmer at the surface of RE projects.

The Mechanics of it: Good Compliance, Bad Development

Lack of dedicated development expertise

All projects, by virtue of their close links with communities, require a community liaison officer (CLO). However, the inexperience of many IPPs in the South African context has resulted in the complete absence of CLOs or the appointment of incompetent individuals who are deemed to be right for the job because, to many, development is science-less fluff. Therefore, many projects suffer from poor community relations owing to two factors: poor communication and a lack of dedicated attention. These two issues are a function of a greater problem: development professionals are few and far between in the sector, and this lack is felt from within the DoE to DFIs and ultimately, at project level.

Some IPPs stand out for their appointment of development professionals. These IPPs tend to form part of larger corporations with shareholders that require triple bottom line accountability. For the most part, however, development professionals are missing in action, which is why relations with communities are, for the most part, strained and ripe for crippling political interference.

Politicised job opportunities

Who gets to work on the construction of a power plant? This question is most loaded in relation to local community members who often comprise the bulk of the unskilled and semi-skilled labour force on RE construction sites. Recognising the difficulties that come with the task of employing locals, many projects have opted to delegate this responsibility to local politicians or local government. The local individuals in charge then 'supply' a group of labourers, which projects tend to hire without much due diligence testing.

What the ED Compliance framework fails to monitor is:

- where and how local employment opportunities are advertised;
- what the selection process is for determining who does/does not get work;
- how transparent the employment processes are with respect to the local community;
- what the process is for getting one's name on a local database for unemployed people (in cases where such databases are relied upon)
    - Is it fair?
- Has the local councillor advertised the work opportunities to all communities within a 50km radius, including those that do not fall into their ward? and
- Has the local councillor positioned the work opportunities as somehow connected to their political party?

These questions are pertinent because these projects are usually situated in communities that are afflicted with high poverty and unemployment. In these contexts, employment is highly politicised, meaning that projects are often drawn into local party politics and are viewed as complicit in local systems of patronage due to their oblivion or outright neglect.

Exclusion of local business sector

The programme understands 'local procurement' to mean 'of South Africa'. What this means then is that, in some cases, truly local enterprises from communities surrounding the project are completely excluded from participating as service providers. This outcome, perfectly compliant, misses the opportunity to stimulate the growth of the local economy. This often occurs because: 'the IPP knows not what the engineering contractor does'.

The structure of RE projects generally results in a complete delineation of duties that sees the engineering contractor do all the subcontracting related to the construction and, later on, operations. Therefore, unless the IPP includes a requirement for community-based service providers in their contract with the engineering firm, the obligation does not exist. Another risk related to this separation of functions is that engineering firms do not always collaborate with IPPs in devising strategies around improving local procurement. The result is that the budget for enterprise development, held by the IPP, is not utilised to convert community-based suppliers into 'procurement ready' vendors. This is explored in further detail at a later stage of this section. Suffice to say that community-based suppliers are excluded because the compliance framework does not require their inclusion and because engineering firms usually lack the understanding to leverage funds from IPPs in order to make local vendors procurement ready.

Much like job creation, there exist, at the local level, databases pertaining to the local business community. Therefore, where projects have identified services that can easily be supplied by local businesses such as the washing of solar panels, they might seek out community-based businesses. However, it is often the case that contractors rely on local politicians or local government to avoid the responsibility of directly interfacing with the communities they are located in. The result is that contracts are then awarded to entities that are connected to the local political power structures.

Socio-economic development during construction

One of the most critical risk factors inherent to the design of the programme is the timing of social development investments. The programme is designed to enable IPPs to direct funds towards development during the operations phase of power plants. The logic is that, at that point, the plant is generating income from the supply of power and is therefore able to free up revenue for socio-economic and enterprise development.

What this timetable has not fully understood is that the construction of mega projects, often in very poor communities, indicates from the very first day that there is big money in the air. So, naturally, projects have found themselves under pressure from communities that insist that there be some level of social investment during the construction phase. Why does this happen? It happens because the current design of REIPPP has at its heart a notion that says it is reasonable to ask of IPPs to risk millions on physical infrastructure, but somehow it is unfair to ask of those same IPPs to invest, at risk, in the communities they are located in. This is rooted in a failure to articulate the return that can be realised from positive social relations. Indeed, experience tells us that, just by completing the development equation with respect to jobs and procurement, that is by investing in skills and supplier development, IPPs would demonstrate high levels of goodwill and avoid pressures that see them sponsoring random social projects under duress.

Location! Location! Location!

The current framework is designed to incentivise projects to focus their efforts on the communities closest to them. Project sites are typically chosen on the basis of access to land, proximity from and ease of connection to a sub-station, ease of access to national roads and general ease of land use such as minimal interference with natural life. Because of these requirements, projects are typically located in areas with low populations. Therefore, a project that commits to investing its socio-economic and enterprise development contributions within a 50km radius achieves more points than a project investing the same amount beyond the 50km radius. The natural outcome is that all projects commit to investing all their social development revenues within the 50km radius because what they are after are maximum compliance points, which is not the same thing as maximum development impact.

The real nature of the development problem with respect to where investments are made relates to three issues.

The first is that, at times, projects are located in areas so remote that there are less than 5,000 people living within the 50km radius. This means, for example, that over a 20-year period, less than 5,000 people will be the recipients of up to R100 million (a conservative estimate of dividend earnings due to a community with a shareholding in power plants), which would imply an over-saturation of investments that could have a higher impact if spread amongst more communities.

The second issue is that working within the 50km radius to achieve maximum compliance can also result in the construction of false borders within related communities. This is not unlike the straight lines that cut up Africa into nation states: straight lines that create unnecessary bureaucratic complexity and impose a regime of inclusion and exclusion in the name of compliance points.

The final issue is that complying with the 50km radius is also a recipe for generating high levels of inequality within neighbouring communities. This is compounded by situations where two power plants are built right next to each other making a single community the recipient of benefits from both. This inequality has the potential for a myriad problems related to falsification of identity, the influx of the excluded group into the suddenly wealthy community and the stirring of tensions and resentment related to what is a highly artificial premise for accessing benefit.

How Does This Compromise Development?

The real and ultimate outcome of not appreciating the risks entailed in poor compliance design and the resultant development malpractices are life-threatening community protests that might include the possibility of power plants being set alight by angry communities. The irony is that, while projects are willing to take into account the possibility of the sun not shining for protracted periods despite millennia of experience to the contrary, the possibility of community-level protests as a result of development malpractice is not taken seriously, despite South Africa's history as well as the current trend of violent community protests for improved service delivery.

The implication is that all actors involved are risking these energy assets. Because development is not understood, and few are willing to appoint dedicated professionals to this function, the short-term risk is reputational but, in the long-term, we risk catalysing sustained community development where funds would otherwise not exist. Turley and Perera make a similar point in their assessment of sustainable procurement. They identify the resourcing of municipalities with sustainable development professionals as a key component of identifying, mitigating and managing the social risks that emerge from such investments.[37]

Future Research Implications

Part of the reason that the return to social investment is undervalued is because there exists little to no research about social risk: what it is, how to measure it, how to manage it and how to realise returns to investments. This research is necessary to provide both evidence and cases around which to base social investment choices.

Hurdle 2: Treating Communities as the Sole Representative of Black South Africans

The current design of the procurement programme makes it possible to treat a community, represented through a community trust, as the sole representative of Black South Africans in the ownership structure of IPPs. This is not inherently problematic, but can only be an appropriate strategy in cases where individuals who have the fund-raising and deal-making

37. Turley, L. and Perera, O. 2014. 'Implementing Sustainable Public Procurement in South Africa: Where to Start?' IISD Report.

experience to negotiate on behalf of their communities represent the community trust in question.

The Mechanics of it: Good Compliance, Bad Development

In reality:

- Most community trusts are formed for the purpose of a specific bid. In these cases, the representatives of the community are usually new to their roles as board members or trustees and bring no personal experience in terms of fund-raising or engineering.
- Those trusts that have been in existence for longer are often products of previous bidding rounds and have not begun to operate because their trusts have not received dividends and, therefore, there have been no decisions to make. This means that, once more, the community members who represent the community on the trust's board generally have limited relevant experience.
- Because IPPs are allowed to elect their own representatives to the trusts' boards, these individuals, representing the IPPs interests, make all key decisions regarding how the community is to be funded and what the focus of the community trust should be.
- Furthermore, this means that the IPP, as a project company, can in fact be comprised of a single entity, effectively negotiating with itself due to the inability of community trustees to participate.
- Therefore, where South Africans and Black people are intended to participate, it is entirely possible (and has been the case) that foreign-owned companies can construct an entire bid and be deemed successful from a compliance point of view, without any meaningful contribution from local parties.

How Does This Compromise Development?

Ownership as a factor of production is rewarded through profits. However, the rewards that accrue to ownership, which make it possible to reproduce profits over time, are a function of the capability to organise capital and all other factors of production towards a single purpose. This means that, to convert the element of ownership into a capability that can be reproduced by South Africans, it needs to be approached not just as a question of the outcome: that is, whether or not a Black person receives a dividend when profits are declared. Instead, ownership needs to be approached as a question of capabilities, which are expressed in multiple processes, and life stages of an

IPP: from project development to fund-raising to deal-making and, crucially, to operational involvement in the core business of the IPP. This means that the DoE should request substantiating information that details the involvement of South Africans in general, and Black South African owners in particular, in all phases of the IPP's life.

Where community trusts are concerned, there should be an awareness of the limitations of trustees with a view to developing their capabilities over time. Instead, the strategy of choice is to give community trusts economic interest without decision-making power, which is viewed as a way of protecting them from the full implications of the fiduciary duties that would accrue to trustees. But this is, at best, a stopgap. It cannot be treated as something that will self-correct over time in the absence of any intervention. Rather, the long-term strategy should be to train trustees so that they may play a meaningful role, not just in community-level issues, but also in the management of the power plants owned by their communities. This is not a requirement of the programme thus far, which explains why some projects find the assignment of Black South African ownership entirely to communities to be a good strategy because it guarantees that one half of the ownership equation will always be unable to negotiate or vet the decisions of the other. What the DoE and development financiers should be asking is what the strategy is to ensure that currently passive community trusts can grow into genuine shareholders with the ability to negotiate their interests and, ultimately, to continue the work of the trusts in the absence of these projects. This is a question of empowerment and a question of sustainability. Failure to answer it is inadvertent complicity with a long-term tokenistic role for communities as shareholders represented by community trusts.

Future Research Implications

The key research questions that stem from the capabilities question are related to prioritisation. What skills are most pertinent to the active participation of Black South Africans? What vehicles would be the most effective in imparting those skills? And, in the interim, what services must be availed to participating entities that are still in the early stages of skills development?

Hurdle 3: The Problem of Absent Black South African Owners

Related to the point above is the challenge of absenteeism where Black owners are concerned. One unfortunate by-product of Black Economic

Empowerment (BEE) is that it has created a schism between the Black firms that are created solely for the purpose of investment holding and the firms that are created for the purpose of operating the assets they own. There are, of course, some great examples of firms that have a core operational competency as well as the ability to raise funding for large investments such as are required for RE projects, but these are in the minority. To address this, the current procurement programme recognises ownership on two levels: the level of the IPP and the level of its most immediate contractors responsible for construction and operations. Therefore, in the ideal world, the Black entities with fund-raising experience apply their trade at the level of IPP ownership and the entities with engineering capabilities can apply their trade in the construction and operations of power plants. There is also a view that the requirement to have Black people in top management positions takes care of the skills transfer question at the highest level. Not so simple.

The Mechanics of it: Good Compliance, Bad Development

- As already indicated, at the level of IPPs, the Black owners who participate are typically skilled financiers. However, they only concern themselves with the financial return profile of the project and do not get involved in the operations.
- Unfortunately, it is also the case that the Black ownership in construction and operations companies is comprised of passive investors who are not actually involved in the management or operations of those firms.
- Therefore, the assumption that Black owners of construction/operations companies can become recipients of the technical capabilities entailed in designing and managing renewable energy power plants is erroneous.
- In effect then, the entrepreneurial capabilities that are required to create and maintain an RE power plant, from fund-raising to engineering, are not being transferred to whole Black South African entities that have that complete set of skills.
- If we are to accept that we do not require a single entity with all the key competencies that comprise an IPP, then we might find comfort in the existence of the requirement for Black top managers. But here too exists a challenge. From a compliance perspective, a project can claim to have only one or two top managers, all Black and female. In this way, the project is awarded full compliance points. That these two individuals may be focused on non-core issues is not something the DOE monitors.

Even more concerning is the fact that there may be a host of foreign individuals who occupy the key roles in the top management of an IPP who are completely invisible to the DoE. This is because the structure of a top management team is completely discretionary and therefore makes it possible to conceal the reality of the limited role of Black top management in the actual running of IPPs. Per the compliance obligations, such a concealment of the reality is not a contravention and, despite agreement that this sort of behaviour goes against the spirit of the DoE's intentions, there is no way to reward or punish IPPs, because the DoE has not explicitly set out to measure whether or not Black top managers are at the heart of executive decision-making.

How Does This Compromise Development?

The above is a problem for development because it means we have so far failed to create a complete entrepreneurial class for this sector. Instead, it is possible to be compliant with a structure that reinforces passive Black involvement, confining this class to the role of dividends collectors or, indeed, highly paid Black top managers whose functions are mostly ceremonial. Rather, what is needed is a level of detail regarding the operational roles of Black owners in IPPs, construction companies and full top management structures that indicate what Black top managers actually do in relation to their counterparts in executive management. This should be interrogated in the assessments of bids and monitored throughout the life of the power plant.

Firstly, this approach will avoid the problem of token Blacks or South Africans. Secondly, it will avoid the problem of foreign IPPs subverting their responsibility to partner in a mutually beneficial way with local entities. Thirdly, and most importantly, it will result in what is actually required: the creation of a class of Black or South African industrialists who can single-handedly create renewable energy IPPs in the future. Indeed, there will be challenges regarding the experience and capital that Black or South African entities with such aspirations can currently bring to the fore, but the benefit of the last 20 years is that, despite the shortcomings of BBBEE, many operational Black entities have emerged who, if unable to bring a cheque, can bring a myriad capabilities that are required to run an IPP. Where there are gaps in the experience of such entities, these should be identified in the Detailed Economic Development Reports of bidding IPPs, in tandem with a development strategy to overcome them.

Future Research Implications

'Local ownership may be desirable, but it is not the same thing as capacity building, which involves the development of managerial, technical and operational skills in national firms and the domestic labour force. And to achieve the highest level of industrial capacity building, policymakers must focus in a coordinated way on basic policy deficiencies affecting infrastructure development, trade/industrial policy, and skills development and transfer.'[38]

The current call for Black industrialists indicates the state's awareness of this challenge at the national level. What is of interest is research that explores enterprises that have participated in BBBEE deals, the skills they had going into the deals and the ways in which they have subsequently developed. It may be the case that skills do in fact trickle down through a less regulated manner than what is proposed. Of even greater interest is how this process of up-skilling occurs and what incentives drive the parties involved in giving and receiving.

Hurdle 4: Jobs without Skills Development

The job creation impact of RE projects is monitored very closely. The DoE concerns itself with who is employed, their nationality, race and gender, the employment of local community members and the identities of the skilled workforce.

One of the job creation challenges faced by the RE sector is the limited time period in which power plants can actually generate meaningful, mass employment. Typically, an RE power plant is a fully automated generation facility that does not require person-power to convert its energy source into electricity. Therefore, labour is most pertinent to the construction phase, which typically takes 9 to 36 months.

Thereafter, the plant does not require a large workforce as the bulk of activities related to its maintenance include activities such as washing solar panels, landscaping and security.

To give an indication, Eberhard, *et al.* demonstrate that the last three rounds of REIPPP have produced 64 successful projects which will result in the generation of 3,915MW of energy.[39] Of those projects, two are hydro, one is landfill gas and the rest are a combination of the dominant technologies:

38. Eberhard, *et al.* 2014. 'South Africa's Renewable Energy IPP Procurement Program: Success Factors and Lessons', PPIAF.
39. Eberhard, *et al.* 2014. 'South Africa's Renewable Energy IPP Procurement Program: Success Factors and Lessons', PPIAF p. 14.

wind, solar PV and CSP. The jobs created by these projects combined during construction are estimated at 19,108 per the DoE's reporting.[40]

A job, according to REIPPP, is calculated as 12 months of full-time employment, measured in terms of time worked rather than the number of people who are engaged. The REIPPP Request for Proposals refers: 'A "Job" is accordingly calculated on the basis of total Person Months for the Construction Measurement Period and the Operating Measurement Period, divided by 12.'[41]

To be clear, per the DoE's definitions:

- A person month is 160 hours of work
- A job is 12 person months

In other words, if one were to work the same job for 10 years as a full-time employee, they would accrue 120 person months, meaning the DoE's statistics would reflect 10 jobs even though only one person has worked that time. To get a sense of how many people are employed then, one has to divide the number of jobs by the duration of employment. Construction periods vary depending on the size of the power plant. It is reasonable to assume an 18-month period as representative of the average period. In that case, we divide 19,108 by 1.5 years in order to get a sense of people employed during construction. This would mean that roughly 12,738 people will have worked on constructing the first three rounds of power plants, which is significant, but severely compromised by the fact that these jobs do not last beyond the two-year mark.

Eberhard, *et al.* further report that the dominant technologies will generate 34,954 jobs. Using the same division principle as above to get a sense of how many people will have jobs for the 20-year operations phase of these projects, we divide the total jobs by the number of years. This gives us a total of 1,747.7 people who will enjoy the benefits of working on power plants during operations. This number lacks some lustre relative to the R120 billion that has been invested in the sector over the first three rounds.[42] This is because this sector is not synonymous with creating long-lasting, quality jobs.

So how, then, can RE projects deepen their employment impact?

40. Eberhard, *et al.* 2014. 'South Africa's Renewable Energy IPP Procurement Program: Success Factors and Lessons', PPIAF p. 27.
41. DoE. 2013. 'RFP – Volume 5 – Economic Development Requirements', Part 4.2.7.
42 Eberhard, *et al.* 2014. 'South Africa's Renewable Energy IPP Procurement Program: Success Factors and Lessons', PPIAF p. 14.

The Mechanics of it: Good Compliance, Bad Development

Currently, the question of employment impact does not feature in the programme's compliance requirements. Therefore, it is the case that people are employed for a short period and are released when their roles are completed without any prospects for future employment. This outcome is completely compliant.

How Does This Compromise Development?

To be fair, it is not the responsibility of IPPs to guarantee the future employment of staff they no longer require. However, a developmental approach to this question would assess employment impact from the perspective of employability. In other words, what IPPs can be measured against is how they improve the future employment prospects of their labour force in the limited time they have with them. The answer is two-fold: power plants must become sites of learning and such learning should be recognised through certification for those who then demonstrate graduation from one skill set to the next. Failure to maximise the limited time with workers simply perpetuates the underlying reasons for their unemployment. It is not unreasonable to request that IPPs design training programmes for workers. The challenge is that the link between job creation and skills development is not an obligation and, therefore, making this investment in workers currently constitutes undue expenditure.

Hurdle 5: Procurement without Supplier Development

Another common problem relates to entities that provide services for RE projects. The programme sets out rules that require IPPs to procure from BBBEE accredited agencies and the value of their invoices is then recognised in terms of their BBBEE level. In this way, the programme rewards procurement-spend on entities that are highly compliant. Included in the compliance framework are the requirements to procure from small enterprises and women-owned vendors.

The Mechanics of it: Good Compliance, Bad Development

There is, however, no obligation to procure from local communities and, furthermore, there exists no obligation to develop the small suppliers from which projects procure. The result is a crude, last-minute approach to procurement, which at its worst results in the creation of small enterprises and women-owned vendors that assist with meeting compliance obligations but are not assisted to develop beyond servicing the power plant.

How Does This Compromise Development?

This represents a missed opportunity for development because, with the correct level of support, small enterprises should be able to leverage the experience of servicing power plants to grow into larger, sustainable businesses: because, indeed, the growth of small businesses, particularly in new industries, is a prerequisite for the growth and sustainability of the economy at large.

Future Research Questions

The acceptance of procurement as a tool for enhancing the economic participation of previously excluded enterprises lacks an evidence basis. Indeed, Tait notes that the failure to review enterprise development has led to a poor understanding of how to ensure that obligations result in sustained benefit.[43] This historical impact, then, is the question at the heart of procurement.

Hurdle 6: Research! Research! Research!

Power plants must submit detailed environmental impact assessments for consideration by the DoE. These are scrutinised and, where there are gaps or risks, IPPs are usually requested to submit mitigation strategies. Where communities are concerned, a document relying on census data and dated Integrated Development Plans (IDP) is often deemed sufficient evidence of a socio-economic needs assessment. And if the content is ever to be interrogated, it will not be by a financier or the DoE but most likely a powerless NERSA official whose contribution can only be to raise a flag at a public hearing. This is how unimportant social impact is in the RE sector.

Simply, the recourse is to insist on participatory research prior to the submission of a bid. Once a licence is granted, the DoE should insist on participatory development planning – a form of development planning that will occur in collaboration with communities not solely in the plush Cape Town or Johannesburg offices of IPPs. Furthermore, the DoE should insist on the monitoring and impact evaluation of social investments, which goes beyond evidence of a money trail and assesses the effectiveness of development interventions.

The problem of poor development planning has been identified by many who have followed the sector closely. Wlokas, *et al.* identify the absence of community development guidelines as central to the challenge, noting that

43. Tait, L. 2014. 'The Potential for Local Community Benefits from Wind Farms in South Africa', Energy Research Centre, University of Cape Town.

IPPs are not experts in development and, therefore, that their limited understanding and resultant negligence are to be expected.[44]

The bottom line is that conducting detailed and inclusive research in communities impacted by RE power plants should be a minimum standard for demonstrating seriousness about achieving the development aims of REIPPP.

Hurdle 7: Community Participation

Here is a proposed definition of a community: a grouping of highly underestimated individuals. Why? Because most actors in the RE sector think it perfectly reasonable to remove the function of thought from community members. After all, poor people are, by definition, uneducated and therefore unable to articulate their needs and aspirations, right? Wrong.

And yet, there is nothing in the compliance framework that explicitly requires evidence that communities have been consulted about how their collective identities are to be used to acquire electricity generation licences. Certainly in the bidding process the environment's interests are even more important than those of the communities. And post the bidding process, a largely ceremonial process run by NERSA is embarked upon to publicise the projects that are likely to be built in communities. These hearings are generally held in towns that, although in the same province as the projects in question, may be over 100km away from the communities impacted by them. In other words, NERSA hearings cannot be relied upon to get the word out to communities or to listen to the views of community members.

There is a deeper problem that underlies the convenient distancing of communities from projects: the fear of raising expectations, which is actually an expression of the inability to manage expectations. The sector is replete with actors who harbour cynicism about communities, lack the flair to manage local community politics, and undervalue constant, transparent communication with ordinary members of communities. This issue is certainly not limited to the sector but it acts as a barrier to development relative to the spirit of REIPPP and the national transformation agenda.

But if communities knew more they would demand, at the very least:

- to be notified that they are beneficiaries of community trusts;
- to be consulted about the social needs that many IPPs claim to respond to;

44. Wlokas, H., *et al.* 2012. 'Challenges for local community development in private sector-led renewable energy projects in South Africa: an evolving approach', *Journal of Energy in Southern Africa*, Vol. 23, No. 4.

- to be notified of the individuals who are elected to represent their interests on community trust boards;
- to be made aware of the job opportunities and selection processes for the construction and maintenance of power plants; and
- accountability with respect to the plans that IPPs submit to the DOE regarding community development.

The unintended effect of the generalised silence where communities are concerned is politicisation of access to information. Those in the know form part of a connected minority who can use the power of their knowledge to become unofficial gatekeepers. It also becomes an instrument for avoiding accountability because the standards by which IPPs are to be judged are conveniently concealed. Some erroneously link this form of gatekeeping to the heritage of traditional leadership. However, the complicity (and sometimes active agency of foreign IPPs in limiting access to information and working with a chosen few in communities) indicates that this behaviour is less inherent and more opportunistic.

Further Research Questions

What is necessary to help bridge the development training divide are short courses in development studies and related disciplines, development for non-development professionals, as it were, to heighten awareness of the issues. Furthermore, research demonstrating which tools are best suited to managing community participation should be conducted and disseminated to create a shared set of minimum standards across the sector.

Hurdle 8: Limited Time as a Guise for Limited Will/Capacity

The standard response to many of the above issues is that there is simply no time to address development in a thorough manner. The power crisis is so urgent that some development concerns must be dealt with at a later stage, would say a DoE official, a well-intentioned IPP or a DFI. However, if we consider the complexity of the technical requirements to build a power plant, the long hard process of raising debt and equity for such an endeavour and all the legal work that currently underpins REIPPP, it would seem will, not time, is the real issue.

Will is governed by depth of understanding and incentives. As it stands, development is not fully understood by the people who hold the power. Some hold the view that development practitioners are naïve tree-huggers with no claim to science or professionalism; other IPPs are cynical about communities and prefer to present their development obligations as non-negotiable acts of charity rather than the entitlement that it is; many, afraid to confront the power dynamics that exist within communities, choose to avoid responsibility by working through local politicians or community workers whose pasts or intentions are not vetted. More often than not, however, it is a lack of understanding that leads IPPs to behave in this manner.

Hurdle 9: Doing is Not Impacting/No Impact Without Investment

There is a common view that, by doing something perceived to be positive for a community, development is achieved. This is the view of those who paint old age homes and place computer labs in schools with intermittent electricity supply. 'We've done it, branded it and have the glossy picture to show for it.' Therefore, development has taken place? Not. This understanding of development is probably why decades of CSI have, in the majority of instances, not translated into the comprehensive upliftment of communities. It is also the reason why one of the most influential development thinkers of our times is Professor Esther Duflo, whose work is premised on questioning the impact of development projects and investments. Doing what we perceive to be good is simply not correlated with achieving development impact. This is why development requires expertise.

Despite this, the expertise of compliance, essentially auditing project activities, is ranked more highly than development impact evaluation in the RE sector. In fact, impact evaluation is not a requirement of the sector, meaning that, for the time being at least, doing is erroneously assumed to be the same thing as impacting.

The additional assumption that pervades the RE sector is that all social development investments can be directed to a capable community trust, whose sole function will be to identify existing community-based organisations and enterprises to which funding is directed. Firstly, as discussed above, most community trusts that represent communities in the shareholding of RE power plants are completely new structures with very limited experience in the functions they are expected to carry out. Secondly, many power plants are located in areas that are extremely under-developed, which by definition implies weak or limited community structures. As evidence of how limited the understanding of communities is, it is commonplace for projects to submit ED Reports that claim their newly established community trusts in communities with few identifiable community-based organisations will handle the investment of millions of rands. This is deemed compliant, despite its complete lack of development logic. Rather, what is required, despite the effort it entails, is the creation of robust local organisations that can implement all development functions at the community level from research to planning, implementation, monitoring and reporting.

As with all the other hurdles explored, in the absence of definition and measures, development impact cannot be guaranteed through the programme as it currently stands.

Hurdle 10: 'You Cannot Manage What You Do Not Measure'

To compound what we have shown to be a limited understanding of development across the sector is the very real issue of the DoE's inability to monitor the development efforts of IPPs. This has led many IPPs to view any investment in development as wasted expenditure because financiers and the DoE do not provide incentives related to the quality of development work. Rationally then, unless the threat of a community uprising looms, many IPPs practise minimum or bad development because, to begin with, they lack a full understanding or appreciation of the returns on good development and, more importantly, their efforts are not recognised or rewarded in any way under the current compliance framework. Indeed, 'you cannot manage what you don't measure.'

In the proposed seven equations of development, what we contend is that the latter part of each equation must be included in the ED framework. This means that it must be assessed as a composite part of the bid; there must be minimum performance levels associated with development, and clear consequences for risking the national fiscus through the active subversion of development.

Rather than further detailing the implications of the failure to measure, we complete this section with a high-level proposal for how development can possibly be measured within the context of the proposed Seven Equations of Development.

Measuring and Managing Development

Development Element	Measure	Phase of Measurement	Evidence of Activity
Fair job creation PLUS skills development	- Number of employees trained - Relevance of newly acquired skills for further employment	Construction and operations	- Training attendance registers - Awarded certification - Democratic access to work and training opportunity
Procurement PLUS supplier development	- Number of local suppliers procured from - Value of procurement spend	Construction and operations	- Investment in procurement readiness programmes - Democratic access to opportunity
Ownership PLUS operational involvement	- Defined operational roles for Black/South African owners measured in terms of type of work and time worked	Construction and operations	- Letter of appointment - ID - Time sheet - Salary advice
Management PLUS key roles for Black/South African managers	Black/South African managers' roles measured in terms of type of work and time worked relative to the full executive team	Construction and operations	- Letter of appointment - ID - Time sheet - Salary advice
Socio-economic and enterprise development PLUS community participation	- Participatory research - Participatory project design - Training and development plan for trustees - Case for beneficiary community selection - Case for trustee selection - Communication strategy	Construction and Operations	Detailed reports - Imbizo minutes - Financial investment records - Development practitioner CV - Communication strategy

The above recommendations are proposed as a starting point for the sector to assist all actors – project developers, IPPs, financiers and the DoE – to assess and monitor the quality of a project with respect to development. In each instance, there should be a clear identification of the implementation agent. While the responsibility lies primarily with the IPPs, financiers and the DoE, these undertakings should be dealt with as a collaborative effort, involving such state agencies as sector education and training agencies and departments such as trade and industry, small business development and higher education.

CONCLUSION: THE TIME IS NOW

This chapter has deliberately confined its critique and suggestions to the current REIPPP framework because, in spite of its design flaws, research and experience have shown that a more sustainable and inclusive form of development is possible through a reconfiguration (rather than a total overhaul) of the existing ED framework. Secondly, we seek to provide actors in this sector with practical ways of thinking about, measuring and implementing development in relation to compliance, hence the table presented in the conclusion to the previous section. A clear choice was made to avoid naming and shaming individual projects because the malpractices that have emerged in the sector are largely owing to the programme's design rather than explicit malicious intent. We believe that practices can and will improve if the correct incentives are put in place. Therefore, it is less the story of individual perpetrators that makes the case and more an understanding of generalised trends that must be internalised in order for genuine change to come about.

We believe that 'The 10 Hurdles' that currently hinder development in the sector can be overcome by measuring and managing the proposed 'Seven Equations of Development'. By applying this thinking to development, the programme can contribute the following to the national development agenda:

- the creation of Black-owned and run energy companies, which also implies the establishment of genuine Black/South African industrialists;
- the transfer of skills from foreigners to locals in the most senior executive functions as well as the most junior roles entailed in constructing and operating RE power plants;
- the creation or growth of small enterprises that can leverage the

experience of servicing power plants to participate in other sectors with similar needs;
- the development of communities into active agents in their own story of 'good change';[45] and
- impactful social investments owing to the participation of local communities in the articulation of development strategies and their resultant stewardship in managing the investments made in their communities.

However, this is not a nice-to-have that is secondary to the work of financing projects and identifying the correct technologies. What this chapter has demonstrated is that failure to apply good development thinking has already resulted in fractured community relations, unsustainable job and enterprise creation and ineffectual social investments made to appease rather than advance communities.

It is thus our recommendation that financiers, the DoE and IPPs should immediately introduce changes that will result in the management of the very real social risks that are currently being allowed to persist. Communities are gradually waking up to what they are truly entitled to and bad development is not it. There is certainly a role for communities to put pressure on IPPs and the DoE and the hope is that it takes on a constructive form rather than the exacerbated infrastructure destruction that has come to typify community protests. Indeed, the House of Traditional Leaders in the Eastern Cape has shown signs of positive engagement on the question of land use by challenging the DoE's rules with respect to traditionally owned land. Their argument is that the rules of REIPPP exclude their participation by requiring that power plants be built on privately owned land, which naturally excludes the communally held, untitled land of Black communities.[46] That such engagement, in the case of communal land areas, should also genuinely involve communities is critical for revealing and addressing the development shortcomings of REIPPP. All this is to say that, since it is not the intention of the programme to compromise its own development objectives, let all social risks that result from it be understood, managed and resolved to ensure that real development is attained and sustained.

45. The term 'good change' is borrowed from Robert Chambers (2004: iii, 1–2), who argues that development is 'good change', thus making the two concepts synonymous.
46. http://www.iol.co.za/business/news/leaders-fight-for-wind-farms-1.1729636#.VQPJ6BCUdfw [Accessed 5 March 2015].

SECTION II

Transition to a Low-carbon Economy

CHAPTER 5

Making Transitions to Clean and Sustainable Energy in the South African Urban Transport Sector: Linkages to Growth and Inclusive Development

Lynn Krieger Mytelka

ABSTRACT

This chapter analyses the development of the bus rapid transit (BRT) systems in Johannesburg (Rea Vaya) and Cape Town (MyCiTi) and the extent to which they have played a role in stimulating a move towards energy transitions. The two cities have each engaged in an effort to build their urban bus transport infrastructure, mainly along the now traditional lines of bus rapid transit (BRT) systems, by the expansion of their bus networks and the purchase of new buses. Looking at past practices, a central part of this analysis is to understand the challenges of this sector and provide policy recommendations on how to overcome the challenges of path dependency and lock-ins associated with the transition to a low carbon-energy system in the transport sector in South Africa.

INTRODUCTION

Since attention turned to the transport sector in the mid-1990s, the focus has largely been on its role in facilitating mobility to and within South Africa's cities. In the ensuing years, considerable progress was made in building the public transport sector itself and in addressing the issue of transport planning 'from a policy perspective' (Bickford, 2013). Nonetheless, as the literature review prepared by the South Africa Cities Network (SACN) noted, a number of new issues have arisen that now require attention.

Several of these issues relate to the recent devolution of responsibilities and functions for the transport sector to the local level and the ability of local authorities to 'become important institutional players in the cities' (Bickford, 2013). Others, triggered by the growing evidence of climate change and its impact (GEA, 2012), take a broader perspective on the transport sector, linking it to the growing need to move towards an energy transition.

The present chapter addresses this new set of issues and the interrelationship between them. Its focus is on the identification of opportunities to begin a move towards energy transitions in the urban transport sector, their linkage to current policy and practice in the development of renewable fuels and their potential impact on innovation and inclusive development.

As part of this process, the chapter analyses the development of the bus rapid transit (BRT) systems and the extent to which they have played a role in stimulating a move towards energy transitions, particularly in the two cities, each of which has engaged in an effort to build its urban bus transport infrastructure mainly along the now traditional lines of the bus rapid transit (BRT) systems. These include Johannesburg (Rea Vaya) and Cape Town (MyCiTi). Expansion of bus networks and the purchase of new buses as part of a second phase is now underway in these two cities and this provides an opportunity to take a new look at past practice and analyse the challenges that lie ahead in moving towards an energy transition in this sector.

UNDERSTANDING TRANSITIONS

There is no universal 'right' pathway to clean and renewable energy. There is widespread recognition, however, that the pace at which such technologies are being developed and diffused is far too slow (Peters, 2011). A consensus is also emerging on the need to stimulate movement in this direction now. Choices about how to proceed will have to be made and new strategies developed and implemented.

South Africa is not alone in facing such challenges. A number of recent studies have sought to analyse the problems associated with energy transitions. Two of these dealt specifically with the process of making choices about change, and they have informed the work to be undertaken here.

The first, through the lens of an innovation systems approach, analysed the growing interest in the application of hydrogen and fuel cells in the transport sector that emerged in the mid-1990s and the abrupt decline in the pursuit of an imminent hydrogen future that took place early in the new decade of the 2000s (Mytelka, 2008, 15–38). It led to an awareness of the extent to which path dependency and lock-in[47] can slow the move towards an energy transition in the transport sector.

The second study combined innovation systems and energy transition perspectives (GEA: 2012, Chapter 25). It made clear the extent to which energy transitions are not isolated events but part of broad, systemic processes of change (Smith, 1994; Grubler, 2004; Geels and Kemp, 2012) during which old and new regimes co-existed, often for long periods of time, and outcomes are consequently unpredictable as we shall see in the brief discussion below on making a late start in moving towards clean technology.

Successful energy transitions, moreover, have historically involved major changes not only in technology but in production processes, management systems, organisational forms, government policies, skills, knowledge (Freeman and Perez, 1988) and, above all, in the habits, practices and norms of the actors involved (Mytelka, 2000). Combined, these processes create powerful self-reinforcing mechanisms that make change very difficult but not impossible. How to overcome the challenges of path dependency and lock-in is a central part of the analysis to be undertaken here.

47. Path dependence is reflected in a range of beliefs and boundaries that shape choices about new technologies. These can include engineering 'beliefs about what is feasible or at least worth attempting', and boundaries that shape processes of choice such as lines of research to pursue, kinds of products to produce, organisational routines and development trajectories to adopt (Teece, 1988; Mytelka, 2008). Path dependence also emerges in contexts where earlier investments result in high sunk costs, habits and practices are entrenched and 'expert views' are shaped by earlier thinking that narrows the range of choices to established technologies and evaluation techniques. See Section I, Chapter 1 for a discussion of these two concepts and the importance that a co-evolutionary role played by governments, civil society and industry can plan in moving forward.

A LATE START

South Africa took a late start in the modernisation of its liquid fuels and it is still in the process of eliminating leaded gas from its roads. In part, this resulted from a lack of sufficient government incentives needed to move in that direction. Unleaded petrol, for example, only became available in South Africa in 1996. The ban on lead in all grades of gasoline only took place in 2006 and the reduction of sulphur levels in diesel fuels is still incomplete.

Although policies in Europe began to focus on cleaner fuels in the transport sector, the process of conversion was slow and the costs high. Moreover, as might be expected in this early phase of a possible energy transition, no consensus has yet been reached among the competing ranks of bus manufacturers in South Africa, each of which appears to have embarked on a different pathway, globally and in South Africa.

Volvo, for example, focused on the electrification of its buses. Having first created a plug-in hybrid, the Volvo 7900, which was available in 2013, it then moved on to develop a plug-in electric hybrid[48] that was launched in 2014. The new bus has an electric motor powered by lithium batteries along with a small diesel engine. It is rapid-charged at selected bus stops. According to Hakan Agnevall, President of the Volvo Bus Corporation, 'Electric-hybrid buses and full-electric buses are tomorrow's solution for urban public transport (Hakan, 2014).'

Daimler, on the other hand, focused on 'cleaner' diesel for its trucks and buses in keeping with EU standards and policies. It also has had a long-term interest in hydrogen fuel cells in the automobile industry. In 2010, with an eye to the future, however, it widened its range and launched its first series-produced electric cars with a hydrogen fuel cell, the new B-Class. Its most recent bus is the Mercedes-Benz Citaro G Blue Tec serial Hybrid that allows emission-free driving under battery power over short distances. The diesel engine in this bus does not act as a primary drive unit, but rather drives the generator to produce electric power as required: this energy is stored by maintenance-free lithium-ion batteries mounted on the roof.

Scania moved in a different direction, announcing a new gas engine for its buses and trucks in September 2010. The gas engines are based on five-cylinder 9.3-litre diesel engine platforms. The engine management system automatically adapts to the gas quality, which provides the opportunity to use different biogas and natural gas mixtures. The gas tanks for buses are installed

48. The initial hybrid was a vehicle with a gas engine and a battery-driven electric motor. The Plug-in Hybrid (PHEV) is an electric vehicle that is part gas and part electric and has plug-in capacity. The newest is an all-electric vehicle that uses no gas and is 100% percent electric.

on the roof. Scania was also involved in the development of clean technology in the bus sector in South Africa, a point to which we will return later.

Unlike the interchangeability of fuels in the flex fuel cars and, more recently, flex fuel buses, introduced in Brazil with considerable success (Grubler, *et al.* 2008), the problem engendered by the development of a multiplicity of new types of buses in South Africa stems from the need to build different types of infrastructure. For cities, the uncertainties of having to make choices from among a variety of new buses that may or may not survive a transition, and do so at a time when older models are still available and less costly, reduces the incentive to move ahead more rapidly towards clean fuels. To a certain extent this appears to have carried over to the first buses purchased for the new bus rapid transit systems, as we will see below.

The slow move towards clean energy has also been affected by the reluctance of the South African Petroleum Industry Association (SAPIA), which, in 2009, argued that 'the move towards cleaner exhaust emissions would require investment of some R40 billion to upgrade existing refineries and change fuel specifications in South Africa' (Njobeni, 2009). Had the process begun then, however, it would likely have been completed by now and at a lesser cost. Instead, the oil and gas industry is still reluctant to move forward towards clean fuels despite their recent report that showed that diesel consumption is still rising in South Africa (Sapia, 13 September 2014). In May 2013, moreover, Avhapfani Tshifularo, Executive Director of SAPIA, speaking at the Fuel Dealers Conference in Johannesburg, stressed that refinery owners were unable to foot the bill and government would thus have to make sure that the 'costs were recovered through the price structure'; that is, by passing the costs onto motorists.

Lastly, government policies are themselves now beginning to have a negative impact on the move towards an energy transition in the public transport sector. This results from the continued focus on cleaner fuels as opposed to stimulating a move towards 'clean' fuels. South Africa's most recent road map, which expired in 2014, for example, was still focused on the application of 'cleaner' fuels by 2017 as opposed to a transition towards clean fuels by that time.

To this must be added the way that policies themselves can generate problems in overcoming path dependence and lock-in. South Africa has a well-developed approach to consultation and a broad set of stakeholders are part of that process. Once negotiated, however, many of these policies are delayed or never carried out. In the initial biofuels policy, for example, South

African fuel producers were to have begun mandatory blending of petrol and diesel with biofuels from 1 October 2015. This has been postponed and it is unclear when it might begin. Small farmers, large farmers, bus manufacturers and members of the business community have all been affected by this practice. More recently, Kobus van Zyl, MD of the Daimler Truck and Bus Groups in South Africa (DTBSA) announced his strong opposition to the delayed introduction of cleaner fuels, which had been planned for July 2017 (*Engineering News*, 4 July 2014), a point to which we will return in the conclusion.

MAKING CHOICES ABOUT CLEAN FUELS IN THE BUS RAPID TRANSIT SYSTEMS

In contrast to the above programmes and practices, which were part of national policies, the development of the bus rapid transit systems was largely in the hands of provincial governments and major cities, which were required to fund their own programmes and make choices about the purchase of buses and fuels. Of particular interest is the extent to which this has opened up opportunities to move towards an energy transition and contribute to inclusive development. This section looks at the first two of the new bus rapid transit systems – Rea Vaya in Johannesburg and MyCiTi in Cape Town.

The Integrated Transport Planning Process (ITP), mandated by the National Land Transport Act of 2001 and its regulations, presented an opportunity for local authorities to address a number of transport issues. At the time, these were mainly focused on transport itself, access to it and the role of public participation in the planning process. The Transition Act of 2003 took this one step further by directing governments to rationalise public transport and decrease the destructive competition between different players, such as the taxi drivers who dominated the urban transport sector in Johannesburg at the time and would later become partners in Rea Vaya.

The first Bus Rapid Transit System feasibility study for the Johannesburg area was carried out in 2006 and in August of that year a fact-finding mission to Bogota, Colombia confirmed their interest in its creation. In November of that year, the Johannesburg City Council approved Phase 1 of the future BRT System and construction began.

In December 2009, a tender for the manufacture and maintenance of 143 buses for Phase 1A of the BRT system was awarded to Scania South Africa for the supply and delivery of the chassis and engines for the buses, while its partner, Marcopolo SA, would supply the bus bodies. Forty-one of the new buses would be articulated buses, which can transport up to 112 passengers. The remaining 102 would be complementary buses that can transport up to 81 passengers.

As Amos Masondo, the then Executive Mayor of Johannesburg, pointed out during a press conference at the time, the buses would all have Euro IV diesel engines with Euro III engines being the minimum standard recommended in the national specification (Masondo, 2009). According to Christoffer Ljungner, Managing Director of Scania South Africa, the city council had set high standards regarding the environmental impact of the bus engines and, after consultation with the Clinton Climate Initiative (CCI), had decided that the engines should meet Euro IV environmental standards of CO_2 emissions. This was one level higher than the existing South African national standard and required a 60 per cent reduction in particulate emissions and a 30 per cent reduction in nitrogen oxide emissions compared to Euro III.

Local government was also conscious of the need to create jobs and Rea Vaya provided an excellent opportunity. From July to September of 2008, 173 jobs, mainly manual, were created. This rose to 467 from October to December 2008; 1,836 from January to March 2009; and 2,803 from April to June 2009. They had estimated that once fully operational, Rea Vaya would offer some 1,500 jobs (*Rea Vaya Joburg*, 2015). For the most part, however, this did not lead to the creation of jobs that were linked to manufacturing, engineering or other production-related activities.

Lastly, although both Scania and Marcopolo had the capacity to manufacture the buses they planned to use in South Africa, the need to ensure that they would be delivered in time for the FIFA World Cup meant that they would have to abandon the production of the buses locally and import them all from their respective companies in Brazil. It was agreed, however, that subsequent orders would be manufactured in South Africa. The first of the complementary buses arrived from Brazil in February of 2009 (*Engineering News*, 6 February 2009). At the end of its first year, Scania had fulfilled its commitment to the Rea Vaya project.

Phase 1 B was launched in Johannesburg in 2013. Mercedes-Benz South Africa (MBSA) was awarded the contract and would work with Marcopolo South Africa who would provide the [bus] bodies. The order covers 134 Euro-compliant Mercedes-Benz chassis that were to be shipped as completely knocked-down down sets (CKC) from Mercedes-Benz Brazil: the bodies were provided by Marcopolo, a superstructure manufacturer with headquarters in Johannesburg. Phase 1C is to start operating in 2017.

As Rea Vaya moved into its fourth year, an opportunity opened up to upgrade the City's older transport system by purchasing between 125 and 175 new buses. In addition to requiring local content, the decision by the City for Metrobus to use biogas follows from a number of studies that have indicated enormous potential for the production of biogas from city waste sources such as landfill sites, sewer plants and grass cuttings by Johannesburg City Parks and the Zoo (Maqekoane, 2013). In addition to providing solid waste for use in the bus sector, organic fraction of municipal solid waste (OFMSW) has a number of unique properties that favour its use. Among these are 'its high efficiency due to its ability to give higher biogas yields of good quality per unit weight than most available substrates and its abundance as well as availability at low cost' (Technology Selection of Biogas Digesters for OFMSW.2014). The City is currently working in partnership with the University of Johannesburg and the South African National Energy Development Institute to refine the technology for biodigesters that can use grass and other non-food agricultural products to produce fuel. From these initiatives one has the impression that Johannesburg is well on the road to clean technology in the transport sector.

Although Cape Town was somewhat slower in developing its Integrated Rapid Transit System, MyCiTi, too, has taken a major leap forward in building the knowledge base for a move towards clean fuels and their linkage to growth and inclusive development. In part this was stimulated by its early link to Scania, which, in addition to its role in the first phase of Rea Vaya, was an early entrant into the development of cleaner energy for buses in Europe. With financial support from the Swedish government, for example, between 2006 and 2009, Scania was able to introduce 140 buses fuelled by a combination of natural gas and biogas into its network in Europe. Since then it has moved onto the development of ethanol buses and trucks, over 1,000 of which are currently in operation in Europe (Stromberg, Director of Sustainable Solutions, Scania).

Support from the Swedish government also provided the funds for a multiyear project in Cape Town that focused on the use of waste as a base to produce ethanol. The biogas used in the buses is produced from organic food waste and wastewater sludge.

Green Cape, a development agency established in the Western Cape in

2010, has been focusing on opportunities to promote manufacturing and employment with a pro-green approach. Two of the problem areas to which they have been paying particular attention are biofuels and the waste economy. Large cities have traditionally suffered from waste disposal problems, but this may be changing and biofuels might have a role to play in this process. Jim Petry, Director of Energy at the Western Cape provincial government, for example, looks at waste streams to produce ethanol and a Stellenbosch University start-up, Stellenbosch Biomass Technologies, began to look for partners in the wine sector in the Western Cape with whom to collaborate. The project would apply second-generation technology to effect the enzymatic conversion of grape skin pomace cellulose into glucose, followed by conventional anaerobic alcoholic fermentation. Insoluble holocellulose and cutin residue from the converted slurry would be turned into pellets as boiler fuel to provide the process energy. Ten rented buses would run on waste from grape skins. In March 2014, Western Cape wine producer, Namaqua Wines, signed an agreement with the Paarl-based distillation design and technology provider, Taurus Distillation, to establish a 1.2-million-litre, 96.4 per cent fuel ethanol pilot plant at its Spruitdrift Cellar, near Vredendal. For this purpose, Namaqua Wines registered Namaqua Fuels and applied for a fuel-manufacturing licence for the latter, in accordance with the requirements of the Petroleum and Liquid Fuels Charter. Namaqua Fuels would eventually blend the fuel ethanol with an additive to make ED95 fuel, which would fuel Scania vehicles fitted with ethanol high-compression engines. The Namaqua Wines and Taurus cellulosic fuel ethanol project was the culmination of an in-depth study on the feasibility of the waste-based production of bioethanol to fuel public transport conducted by the Green Cape Sector Development Agency on behalf of the Swedish International Development Agency, Scania, the Western Cape government and the City of Cape Town (Edited by: Megan van Wyngaardt, 2014).

THE RE-EMERGENCE OF NATURAL GAS

Although both Cape Town and Johannesburg have begun to look into a future that links new energy and job opportunities to the developing bus industry, the pathway to clean energy is far from secure as the current re-emergence of natural gas suggests. Why has natural gas re-emerged at this juncture in the development of clean fuels?

Natural gas is comprised mainly of methane, which has a higher hydrogen-to-carbon ratio than gasoline or diesel and a higher octane number, which makes it possible for engines to operate at a higher compression ratio. Natural gas also has the particular advantage of offering a significant reduction in local air pollutants, especially particulate matter when compared with diesel fuel.

Where natural gas distribution infrastructure is already in place, it is possible to use the gas as a vehicle fuel by installing refuelling systems in grids, for example. This, however, is not often the case in developing countries. Instead, gasoline and diesel vehicles will need to be retrofitted for natural gas, unless they are new vehicles that have been designed for natural gas. Natural gas vehicles can also be biofuel vehicles, running on either gasoline or natural gas or dual-fuel vehicles that run on a mixture of diesel and natural gas. Although the world supply of natural gas substantially exceeds that of oil, natural gas vehicles are more expensive and have a shorter driving range than gasoline vehicles.

The move from gasoline or diesel fuel to natural gas is often portrayed as a positive one that brings us closer to a clean technology. The reality, however, is quite different. Natural gas is still a major greenhouse gas and, although it may be cleaner than coal, it reinforces earlier habits, practices and norms that delay the move towards a transition by suggesting that it can play the role of a bridging fuel until we get something better. Unfortunately, the evidence is beginning to show that it creates a barrier to further change by locking in the new natural gas infrastructure, which has a lifespan of up to 50 years.[49] Nonetheless, on 2 October 2014, Johannesburg Executive Mayor, Parks Tau, in his keynote address at the Rea Vaya Transport summit in Johannesburg, announced the bold plan 'to convert vehicles to run on natural gas ... as the preferred solution to the problem of high oil prices'. The city has 4,000 vehicles and consumes about 19 million litres of fuel per year, excluding the Metrobus fleet (Tau, 2014).

49. Building infrastructure for natural gas, as in the case of fracking, can, for example, chain us to carbon-based energy for as long as 50 years, since the power plants are being built to last that long.

NEW FUEL OPTIONS

Until recently, most biofuels were derived from food crops such as sugarcane, maize and palm oil. These first generation biofuels were alternatively praised as a means to provide energy and reduce greenhouse gas emissions, and criticised for contributing to the destruction of tropical forests and competing with food crops in the use of agricultural land. For the most part, this has been the case, though there has been some acceptance as Brazil has illustrated over the years. Similarly, there is the less well-known example of Jatropha (a plant crop used to produce biofuel), which in some cases has offered an interesting approach to issues of sustainable energy and inclusive development. In 2007, for example, the people of Garalo in Mali, rather than continue to rely on imported diesel fuel for a future off-grid generator, chose to plant Jatropha on 440 ha of their land as part of a multi-goal project to stimulate rural development by providing electricity for lighting, refrigeration, welding, and agricultural processing machinery for use by businesses, health services and schools. Within a few years, the new grid had over 230 paying clients and the farmers, who were active players in the choice process, had not only chosen to plant Jatropha for its oil content, but to intercrop Jatropha and local food crops such as maize, sorghum, millet, peanuts, sesame and beans – thus dealing with the assumed need to choose between food and fuel as some ardent supporters of green technology had earlier argued (Burrell, 2008; Access Sarl, 2010).

First generation biofuels were later joined by a second generation of non-food crops, such as algae and cellulosic materials. In many ways, South Africa is an ideal place to develop algae production. It has heavy industry, the pollution of which could be absorbed as an input into the production of algae along its coastal planes. The key to successful development of algae as a biofuel is to produce the other things that go along with it, since the economics are not all that attractive, which limits commercial involvement. New ways around this, however, are currently underway in the work of local universities, particularly the University of Cape Town and Stellenbosch University.

More recently, a third generation has emerged and has been gaining attention. These have often been called Drop-in Fuels. Unlike many of the earlier efforts to develop clean fuels, the new drop-in biofuels have a number of characteristics that are of particular interest in the broader development of clean fuels. Among these are their interchangeability with existing aviation fuels, which is the leading current interest; their ability to be integrated into existing petroleum fuel infrastructure, which reduces the cost of adoption; and their use of hydrogen, which provides an opportunity for existing refineries to stay alive. One of the most recent feedstock developments has been a tobacco plant hybrid that is growing in South Africa as part of a partnership between South African Airways, Dutch Aviation Biofuels company SkyNRG, and Boeing to 'create jet fuel from Solaris, a nicotine-free tobacco plant' (Kinder, 2014).

Currently, commercial drop-in biofuels, however, account for less than four per cent of the global biofuels market with 'the vast majority of drop-in biofuels derived via the oleo chemical platform which uses feedstocks such as palm oil and used cooking oil, and are often referred to as hydro-treated esters and fatty acids' (Karatzos, *et al.* 2014). These biofuels have been successfully tested in aviation trials but their price remains relatively high.

IS THE FUTURE HYDROGEN?

Unlike the newer biofuels discussed above, current interest in hydrogen fuels and fuel cells dates back to Ballard, a Canadian start-up, based in British Columbia, which developed the first bus powered by hydrogen fuel cells (HFCs) in 1993. Shortly thereafter, they signed a joint venture agreement with Daimler-Benz to work with them on the application of hydrogen fuel cells in the development of their new electric car (NECAR).

By the late 1990s, hydrogen fuel cells appeared to be a proven concept in the future development of clean energy in the transport sector. In particular, proton exchange membrane (PEM) fuel cells used a fuel supply to combine hydrogen and oxygen in an electrochemical process that generated an electric current. The process involves an ion exchange polymer membrane as the electrolyte and electrodes of fine metal mesh on which a platinum catalyst is deposited. It thus converts hydrogen directly into electricity without combustion or moving parts, but the use of platinum continued to make the process costly. Given the interest in clean transport technologies at the time, however, the use of hydrogen in fuel cell vehicles (HFCV), by making them virtually pollution free, was a considerable asset (Mytelka, 2008). An additional advantage for South Africa has, until now, been its abundant supply of platinum.

Since the turn of the century, however, there have been vigorous debates over the future of hydrogen, the lack of available infrastructure for refuelling, and the high cost of operating fuel cell vehicles. Progressively, the time horizon was extended further into the future. Competition, notably from the emergence of new vehicles such as the Toyota Prius, then in its first phase, and growing attention more recently to electric vehicles, and especially hybrid buses and cars, has come to dominate the focus on cleaner technology in the absence of a strong government movement towards clean technology.

Over the last decade, member states of the European Union, the United States of America, Japan and several other countries moved beyond the earlier focus on hydrogen refuelling tests and demonstration projects and several began to test opportunities for making hydrogen fuel cars. The cost of hydrogen vehicles has begun to fall and cities in Europe and elsewhere have built some of the necessary infrastructure. What then has been holding back the move towards the widespread use of hydrogen fuel cells in the transport sector globally?

First, it is important to remember that energy transitions, at least for a time, will involve the coexistence of old and new energy regimes. These are often based on different technologies and energy systems and competition reigns among potential victors. It is thus important to note that hydrogen fuel cells are not the only option available and that lead actors such as Toyota, Volvo, Daimler and Scania are strong players in the race to dominate the change process. Making choices under conditions of uncertainty increases the difficulties in moving to new technologies.

Second, since the development of hybrids, Toyota and its competitors have worked hard to secure the hybrid market and bring it closer to the hydrogen fuel cells of the future. In the past few years, new hybrids have begun to reach the market. These cars, and in some cases, buses, are cross-breeds of one sort or another with various abilities to operate as plug-in electric vehicles. In a few cases, they may even operate as fuel cell vehicles for brief periods of time, while the bulk of the work is done by the more conventional part of the vehicle. Of critical importance is the current lack of consistency across the board and the competition to which it gives rise, without necessarily providing new sources of clean technology.

Third, a move towards the development of new uses for hydrogen fuel cells has recently emerged in South Africa, and is being promoted by existing government and business interests. Both express the current need to widen the use of platinum, which is increasingly under pressure as a key ingredient in the production of hydrogen fuel cells whose cost, though considerably reduced over the years, has not yet reached the level of ordinary cars and buses. Both of these projects are related to the need for electric power, for which there are alternatives as the analysis by Donnell shows (Donnell, 22 May 2015).

Fourth, scientists at several major universities have developed new and promising research of relevance to future hydrogen fuel cells and their use. At Stanford University in California, for example, researchers have developed a cheap way to split water with a single catalyst to separate hydrogen and oxygen continuously. This could be used as a renewable source of clean-burning hydrogen fuel for industrial and transport applications (Cui, 2015). At Cambridge University, scientists have produced hydrogen (H2), a renewable energy source from water using an inexpensive catalyst (Cambridge University research), and the number of similar developments are increasing rapidly.

In sum, changes are taking place but they need to be identified and critically analysed. Some of these are likely to play an important role in shaping the future direction of the hydrogen economy.

CONCLUSION: MOVING TOWARDS AN ENERGY TRANSITION

There is a clear need to begin a process of transition to clean urban transport systems in South Africa: this cannot be done by building inflexible structures based on marginally cleaner bus technologies and energy sources. From the analysis of choices made over the past decade, however, there appears to be little incentive in South Africa to move away from the current approach. Some, but not all, of this may be due to policies or the lack thereof. Governments at all levels, moreover, have moved to support policies of their own choice such as the decision to abandon South Africa's electric car and to encourage a three-year alliance between Eskom and Nissan to test the latter's electric vehicle, the Leaf, in a multi-stakeholder partnership with the support of the departments of Environmental Affairs, Trade and Industry, and Transport and Energy.

Other factors, such as the ups and downs of global pressures and prices are also likely to be of considerable importance in the decisions of local governments to move towards or away from clean fuels in the transport sector. This is particularly striking in the return to natural gas as a clean fuel in the bus sector, as we have seen above.

Within this context, what then might be some of the options to consider in moving from cleaner to clean technologies in the urban bus transport sector and what kind of projects today will introduce the flexibility that might enable movement towards change in the near future? Four of these stand out in particular.

First, it is important to remember that all change processes inevitably face challenges that emerge from established habits, practices, norms and interests, or what we have called path dependence and lock-in. Understanding these habits and practices and learning to work with and around them in a given context is an important first step in an energy transition process.

Overcoming path dependence and lock-in require considerable knowledge about new fuel options as they continue to emerge. In overcoming path dependence on the part of bus manufacturers and related actors, for example, it would thus be critical to know how close to production new fuel options currently are and, in the interim, how interchangeable they would be with existing alternative options as the discussions of Scania and of drop-in fuels both illustrated.

Second, making decisions about moving to new fuels would also require an impartial analysis of the costs that would be incurred and their likely evolution. In the absence of interchangeability, moreover, one might anticipate the need for considerable collective action on the part of governments, users and producers working together. Yet this, as we have seen, is currently elusive where policies that have been announced are not put into practice or have been changed.

Third is the need to reduce the uncertainties that emerge under the moving conditions that characterise energy transitions. Closer linkages between users and producers would be considerably enhanced by greater knowledge about how interchangeable new technologies that are emerging might be and how close these are to full development and use. The change process will also require considerable attention to meeting the concerns of path dependence and lock-in, both of which are enhanced by the weaknesses in carrying out policies and programmes that have been negotiated by stakeholders.

Fourth, and without being overly critical, it must be admitted that the oil and gas industry is still reluctant to move towards clean fuels despite their recent report that shows that diesel consumption is still on the rise in South Africa (SAPIA). Developing the capacity to work with and around such actors would be critical. One of the options is to bring into the choice process linkages between users and producers and their perspective on the nature of existing challenges and how these might be overcome. Closer linkages between users and producers would also be enhanced by greater knowledge of how interchangeable new technologies that are emerging might be, and how close these are to full development and use.

In sum, moving towards an energy transition will thus require the development of new initiatives and a longer-term perspective on new technologies and their ability to be converted into clean technologies over time. This is an important period to understand South Africa's options and develop the strategies that would enable movement towards an energy transition. Currently, as the above illustrates, the various spheres of government seem to be going their own way. That will need to change and new opportunities for working together will have to be created.

References

Access Sarl. 2010. 'Electrification Rurale à Base du Biocarburant le pourghère dans la commune rurale de Garalo dans le Sud du Mali'. Bamako, Mali. Available at: http://access-mali.blogspot.com/2010/09/rural-electrification-of-garalo.html [Accessed 30 March 2015].

ANFAVEA. 2008. *Brazilian Automotive Industry Yearbook*, Sao Paulo and Brasilia, Brazil: ANFAVEA (Brazilian Automotive Industry Association).

Arup, G. B. 2013. 'Literature review on public transport and mobility in municipalities', *South African Cities Network* (SACN).

Burrell, T. 2008. 'Garalo Bagani Yelen, a Jatropha-fueled rural electrification project for 10,000 People in the Commune of Garolo', Mali Folkecenter (MFC), Nyetaa. Available at: www.cleanenergyawards.com [Accessed 25 November, 2013].

Cowen, R. and S. Hultén. 1996. 'Escaping Lock-in: the Case of the Electric Vehicle–Technology Forecasting and Social Change', University of Western Ontario and Stockholm School of Economics. Available at: www.ehcar.net [Accessed 20 March 2015].

Cui, Y. 2015. 'Stanford Researchers Develop Water Splitter for Production of Clean-burning Hydrogen Fuel'. Available at: www.azocleantech.com [Accessed 20 July 2015].

de Bruyn, Chanel. 2009. 'Scania, Marcopolo win BRT supply Contracts'. *Engineering News*.

Donnelly, L. 2015. 'Fuel cell technology could power SA Inc.', *Mail & Guardian*, South Africa. Available at: http://mg.co.za/article/2015-05-21-fuel-cell-technology-could-power-sa-inc [Accessed 30 May 2015].

Edkins, M., Marquard, A. and Winkler, H. 2010. 'South Africa's renewable energy policy roadmaps', for the United Nations Environment Programme Research Project, 'Enhancing information for renewable energy technology deployment in Brazil, China and South Africa'. Cape Town: Energy Research Centre, University of Cape Town.

Available at: http://www.erc.uct.ac.za/Research/publications/10Edkinesetal-Renewables_roadmaps.pdf

Geels, F. W., Dudley, G., Lyons, G. and R. Kemp. 2012. 'The transition perspective as a new perspective for road mobility study', in *Automobility in Transition? A Socio-technical Analysis of Sustainable Transport*. London, UK: Routledge.

Global Energy Assessment (GEA). 2012. Cambridge, New York, Melbourne, Madrid, Cape Town, Singapore, São Paulo, Delhi, Mexico City: Cambridge University Press. Report available at www.globalenergyassessment.org [Accessed 15 March 2015].

Hakan Agneval. 2014. 'Volvo Buses and ABB electro mobility cooperation', *Cision News*. Available at www.news.cision.com [Accessed 30 March 2015].

Karatzos, S., McMillan, J. D. and Saddler, J. N. 2014. 'The Potential and Challenges of Drop-in Biofuels', *IEA Bioenergy Task 39 Summary*.

Kigozi, R., Aboyade, A. O. and Muzenda, E. 2014. 'Technology Selection of Biogas Digesters for OFMSW via Multi-criteria Decision Analysis', South African National Energy Development Institute (SANEDI).

Kinder, J. 2014. 'Aviation Biofuels: Cleared for Take Off', *The Atlantic*. Available at: http://www.theatlantic.com/sponsored/boeing-whats-next-2014/fuels/156/ [Accessed 15 March 2015].

Lakadamyali, F. 2012. 'Scientists produce H2 for fuel cells using an inexpensive catalyst under real-world conditions', *Science Daily*. Available at: http://www.sciencedaily.com/releases/2012/08/120823112927.htm [Accessed 15 February 2015].

Maqekoane, L. 2013. 'Green Buses Drive the City', *Rea Vaya Joburg*.

Masondo, A. 2009. 'Scania, Marcopolo win BRT supply contract', *Engineering News*. Available at: http://www.engineeringnews.co.za/article/scania-marcopolo-win-brt-supply-contracts-2009-01-28 [Accessed 30 March 2015].

Mehlomakulu, B. 2005. 'Hydrogen and fuel-cell technology issues for South Africa: The Emerging debate' in Mytelka, L. K. and G. Boyle (eds.), 2008, *Making Choices About Hydrogen: Transport Issues for Developing Countries*, United Nations University Press, Tokyo and International Development Research Centre, Ottawa, pp.324–345.

Mytelka, L. K. 2000. 'Local Systems of Innovation in a Globalized World Economy', *Industry and Innovation*, 7(1) 15–32.

Mytelka, L. K. 2008. 'Hydrogen Fuel Cells and Alternatives in the Transport Sector: A Framework for Analysis' in *Making Choices about Hydrogen: Transport Issues for Developing Countries* (Editors: Lynn Mytelka and Grant Boyle). Tokyo: UNU Press and Ottawa: IDRC Press, pp. 5–38.

Njobeni, S. 2009. 'South Africa: Switch to Clean Fuels To Cost SA R40 Billion', *Business Day*, South Africa. Available at: http://allafrica.com/stories/200902270102.html [Accessed 15 February 2015].

Parks, T. 2014. 'Natural Gas is Viable'. *Rea Vaya*, Joburg.

Peters, D. 2011. 'Roadmap Towards Cleaner Fuels in South Africa by 2017', Tshwarisano LFB Investment (Pty) Ltd. Available at: http://www.tshwarisano.co.za/index.php?option=com_content&view=article&id=64:roadmap-towards-cleaner-fuels-in-south-africa-by-2017&catid=49:2011-articles&Itemid=64 [Accessed 30 May 2015].

Rea Vaya News. 2015. 'Rea Vaya Boosts Local Job Creation'. Available at: www.reaveya.org.za [Accessed 30 March 2015].

Shankleman, J. 2014. 'Volvo Buses Officially Launches the All-new Plug-in Electric Hybrid Bus'. Available at: http://greenbigtruck.com/2014/10/volvo-buses-officially-launches-the-all-new-plug-in-electric-hybrid-bus/ [Accessed 30 May 2015].

Teece, D., Dosi, G., Freeman, C., Nelson, R., Silverberg, G. and Soete, L. 1988. 'Technological change and the nature of the firm', in *Technical Change and Economic Theory*, pp. 256–281. London, UK: Pinter.

Van Wyngaardt, M. 2014. 'Wine Producers to Establish Ethanol Fuel Plant in Western Cape', Tshwarisano LFB Investment (Pty) Ltd. Available at: http://www.engineeringnews.co.za/article/wine-producer-to-establish-ethanol-fuel-plant-in-western-cape-2014-03-21 [Accessed 15 March 2015].

Vaz, E. and C. Venter. 2012. 'The Effectiveness of Bus Rapid Transit as Part of a Poverty-Reduction Strategy: Some Early Impacts in Johannesburg'. Available at: http://repository.up.ac.za/bitstream/handle/2263/20221/Vaz_Effectiveness(2012).pdf?sequence=3 [Accessed 30 May 2015].

CHAPTER 6

'Green' Policymaking and Implementation at City-level: Lessons From Efforts to Promote Commuter Cycling in Johannesburg

A Discourse on Sustainable Development, the Green Economy and Climate Change

Simone Haysom

ABSTRACT

While central government action to mitigate climate change stagnates and national targets remain woefully unmet, government policy remains critical in addressing the challenge. Many international networks and policy experts posit the role of cities and other sub-national government entities as a partial workaround to this problem. Yet, the majority of city-level 'solutions' are drawn from examples taken from the North. This chapter fleshes out the limits and opportunities of working at city level in a context like South Africa, where many basic development goals have not yet been met.

A case study of efforts to promote commuter cycling in the City of Johannesburg explores the efforts to promote a 'green' goal by local government against the backdrop of highly unequal access to mobility and a very underdeveloped mass transit system. It analyses what has fed into impressive momentum, and what might determine the success or failure of current efforts. Its findings point to the important role that has been played by actors outside of government, such as cycling advocacy lobbies and the universities, but ultimately argues that current efforts need to now focus on addressing equity issues related to making bicycle access and the integration of cyclists in outlying townships into mass transit systems.

INTRODUCTION

At a recent conference in Pretoria on 'Non-Motorised Transport' (NMT), officials from the City of Johannesburg Department of Transport declared their commitment to making Johannesburg 'the most cycle friendly city in South Africa'.[50] This enthusiasm for commuter cycling is borne out in a range of plans and commitments to promote cycling – from a five-year commitment to progressively expand bicycle lanes, to behavioural change campaigns, to preventing cars from entering a large and vital business district for an entire month as part of an eco-mobility festival to be held in 2015. A myriad smaller metros across Gauteng are also accelerating their plans to promote non-motorised transport, as are the cities of Durban and Cape Town.[51]

Currently, South Africa has ambitious national targets for mitigating climate change with little concrete action to achieve them. Likewise, there is little direction from national government on promoting cycling, but rather a national policy that focuses efforts on rural areas. Understanding the factors that have driven the development of progressive cycling policy at city level and ensured its implementation sheds light on how to move beyond the current climate change impasse, and might also suggest how to 'ground' national policies at local level in a hypothetical 'carbon controlled' future.

This chapter aims to contribute to our understanding of the multidimensional nature of 'green' policymaking and policy implementation by examining how one city has gone beyond national leadership. In so doing, it explores the limitations and opportunities of trying to achieve 'green' goals through city-level policymaking.

50. Remarks made at the Second National Non-Motorised Transport Conference Convened at Ditsong National Museum of Cultural History, organised by the Department of Environmental Affairs and Department of Transport, Pretoria, 31 October 2014.

51. See proceedings of the Second National Non-Motorised Transport Conference Convened at Ditsong National Museum of Cultural History, organised by the Department of Environmental Affairs and Department of Transport, Pretoria, 31 October 2014.

METHODOLOGY

Eighteen semi-structured interviews were conducted with a range of activists and lobbyists, including those from the Johannesburg Urban Cyclists Association (JUCA); Critical Mass; the Monthly Cycles and the organising team of the 'Freedom Ride'; consultants working closely with local government, as well as academics, urban planners, and former and current City of Johannesburg employees from the Department of Planning and Development and the Department of Transport over a period of several months between March and September 2014. Informants were specifically asked to comment on issues related to equity, processes of policymaking and attempts to influence policy, the influence of international networks, and how decision-making power was mediated through individual staff in the City of Johannesburg. These interviews fed into a qualitative analysis underpinned by a review of the relevant literature.

Observations were also undertaken through participation in commuter cycling advocacy events and commuter cycling community recreational activities, as well as a conference organised by the Department of Environmental Affairs and the Premier of Gauteng. Data collected by the Gauteng City Region Observatory was used to contextualise findings.

THE ROLE OF CITIES IN MEETING 'GREEN ECONOMY' TARGETS

Cities are an increasingly prominent and important locus of action towards low-carbon societies. Though their contributions are unlikely to replace national and international commitments to reduce emissions, they can play a range of roles in fostering a transition to a low-carbon society (UN Habitat, 2011), for example, through creating incentives in the local research and development industry, legislating against bad practice, adopting cleaner technologies into municipal services and adopting growth strategies that favour lower-carbon systems. Individual and collective actions taken by cities can have significant results and provide a stepping-stone towards rules-based international actions to reduce and adapt to climate change (Goldin, 2013). Higher population densities in cities mean that large numbers of people can be reached through regulation as well as greater efficiencies in rolling out new technologies. Municipal governments are also better able to bring together diverse stakeholders in partnerships to tackle climate change initiatives from multiple perspectives (UN Habitat, 2011). This critical role as 'interface' between different groups in society is particularly important in democratic societies where policies must be introduced to and accepted by different groups in society if they are to prove sustainable and effective.

Cities have become important climate change actors because many governance decisions can and have been made at city scale that are more progressive than international and even national debates. In the US, cities such as New York, San Francisco and the City of Austin have made commitments to reduce emissions that far outstrip the actions taken by national government (Kennan, 2014). Lutsey and Sperling have even argued that the realisation of sub-national commitments in the US in 2008 would make a tangible and significant impact on the national level of Greenhouse Gas (GHG) emissions (Lutsey and Sperling, 2008). In Latin America, cities and their 'visionary' mayors, such as Curitiba in Brazil and Bogota in Colombia, have begun standard examples of cutting-edge urban thinking in talk shops across the world (Greenfield, 2014; Jacobs, 2013; Seijas, n.d.). Across the world, mayors increasingly choose to peg their legacy on projects to 'green' their cities and are increasingly rewarded for it with accolades and press attention (Green, 2013). For example, during his tenure as London Mayor, Boris Johnson publicly set out to 'revolutionise' cycling in London, investing in more prominent cycle lanes and in large-scale efforts to make cycling safer, including through training lorry drivers and improving traffic signage for drivers and cyclists at dangerous junctions, and with recent commitments to increase integration with the underground metro network (Walker, 2013). In New York, former mayor Michael Bloomberg has been credited with constructing new bike lanes, improving old ones, and introducing a successful short-term bicycle rental scheme called Citibikes, lending valuable support to a growing commuter cycling phenomenon (*The Economist*, 2013).

CONTEXTUALISING THE ROLE OF CITIES IN SOUTH AFRICA

South Africa already shows promising signs of city-level initiatives in tackling climate change concerns. Despite little guidance or leadership from the national level in operationalising goals to mitigate or adapt to climate change, several South African cities have drawn up and begun to implement climate adaptation plans, and there is a growing body of literature that addresses the particular dynamics that adaptation planning has taken

(Aylett, 2010; Mukheibir and Ziervogel, 2007; Roberts and O'Donoghue, 2013; Taylor, *et al.* 2014).

Less work has been done on mitigation. To truly maximise the amount that can be achieved at city-level, we need to better understand the structures and cultures of local government and the dynamics of urban politics that are specific to South Africa and how these affect policy development and implementation at the local level.

Forming partnerships and pushing an agenda related to sustainability goals in cities is by no means easy in South Africa. There is a high level of contestation over urban services, low levels of awareness and popular animation about sustainability issues, and the government is still locked in a large quantitative push to deliver goods such as houses in order to redress high levels of inequality. The South African social contract also largely understands development in marginalised areas as taking the same form as the highly consumptive, highly environmentally destructive lifestyles in largely white middle-class suburbs rather than alternative development paths that would emphasise, for example, compact neighbourhoods and mixed-use development (Taylor, *et al.* 2014).

Issues of equity and democratic deliberation complicate efforts to suggest alternative routes of development. How do cities justify adopting more environmentally sound technologies and practices, which are often more expensive than 'business as usual' in the short-to medium-term, when for many citizens basic needs such as access to sanitation and housing have not yet been met? How can negotiations around these issues between different interests be effectively channelled into democratic processes when issues of urban service delivery are so fractious and citizens often fail to ensure accountability?

It is also not clear how to apply locally many of the prescriptions that seem to emerge from the international literature. For example, the role of 'policy entrepreneurs', almost always in the form of an ambitious mayor, are often held to be central to galvanising action (Bulkeley and Schroeder, 2008; Gore and Robinson, 2009).

In South Africa, mayors have yet to feature prominently as political actors, although there is increasingly attention given to the role of local government leaders. Cape Town is the only major city to have been under control of the Democratic Alliance (DA), a party that is the primary opposition to the ruling party in the national parliament. This has meant that, to a certain extent, it has been presented as a showcase for the results of DA government, something that has highly politicised its administration and drawn a lot of attention to the city's leaders (Phakathi, 2014). The growing political importance of local government can also be seen in the rise of municipal service delivery protests (Municipal IQ, 2014). In smaller metros, it has often been local councillors who have received the brunt of criticism for the lack of, or poor quality of, urban services. But these developments have not yet been analysed and framed in a way that highlights where local government decision-making power is really located, or how it is most effectively lobbied.

The influence of international networks also looms large in narratives of city-level change. Several of the large South African cities are indeed enthusiastic participants in global networks that bring together cities under the aegis of peer-to-peer learning and the showcasing of best practice amongst members. These networks include C40 Cities Climate Leadership Group (C40) (Johannesburg is on the Steering Committee and Cape Town is an observer city) and Local Governments for Sustainability (ICLEI) (19 South African municipalities are members). Cape Town and Durban are seen as leaders amongst developing country urban areas in climate change adaptation and their efforts are frequently referenced, for example, by UN Habitat (See UN Habitat, 2011). But these networks are often referenced as evidence of momentum rather than interrogated, with the exception of Roberts and O'Donoghue (2013), whose experience in Durban exposes that the value added by participation in these international processes is by no means straightforward or automatic.

These themes are picked up below when the development of cycling policy in Johannesburg is analysed.

Challenges Facing Low-Carbon Transport Transition in Johannesburg

In South Africa, the energy sector is the largest contributor to CO_2 emissions, having contributed around 89 per cent of emissions between 2000 and 2010 (DEA, 2013). Transport alone accounts for 10 per cent of emissions (DEA, 2013). Addressing this requires several interventions in the transport sector, including a move towards cleaner fuels, as discussed by Lynn Mytelka in Chapter 5 of this book, and moves to both expand the public transport system and discourage carbon-intensive forms of travel.

Cycling fits into the mitigation landscape as it presents a carbon-free mode of commuting, or other trips. Breaking down the carbon contribution of individual modal choices with the transport sector reveals that 19 per cent is accounted for by private vehicle trips and 11.5 per cent by trips on minibus-taxis (SANC, 2011).

However, transport planning in Johannesburg contends with a sprawling, low-density spatial form that arose as a result of apartheid-era urban planning that sought to keep the black working class out of cities or at least far from the urban city centre. This legacy has been further entrenched by housing and land policy in the post-1994 era, which concentrated investment in public housing on the city outskirts, despite the fact that employment opportunities are concentrated in one or more central nodes (Venter, 2011). This means that many people in Johannesburg travel too far to reach their place of work for bicycle commuting to be feasible for the entire journey.

This militates against cycling ever becoming an important primary mode of travel, although it could play an important role as a feeder mode for public transport systems, and also replace private vehicle use for some short trips. It would also need to replace, primarily, private vehicle use, and secondarily, public transport use, in order to have an impact on greenhouse gas emissions. As such it should be seen as an intervention that plays a relatively small role in a transition to low-carbon development on its own, but is useful in tandem with the expansion of reliable, affordable and low-carbon public transport systems.

CASE STUDY: CREATING COMMUTER CYCLISTS IN JOHANNESBURG

Problem and Context

Very little is known about the existing population of bicycle commuters in Johannesburg, including granular data about which trips they choose to make by bicycle or how far their commutes are. This population is in any case small: in 2013, only 0.4 per cent of all trips in Johannesburg were made by bicycle (Wray and Gotz, 2014). NMT – mixing cycling and walking – is the main mode of transport for eight per cent of the population, but this is made up primarily of walkers. A recent 'quality of life' survey conducted by the GCRO revealed that 'many NMT users have below average quality of life compared to other transport users – a finding which challenges an emerging international argument that NMT users tend to have higher quality of life in urban contexts' (Culwick, 2014). The same survey revealed that 46 per cent of cyclists (from a statistically inadequate sample) believe their infrastructure is below average or poor (Culwick, 2014). People who use NMT tend be poorer on average than people who take public transport, and much poorer than private car users (Culwick, 2014). This indicates that walking, and perhaps cycling too, is primarily something people do out of necessity rather than choice at present. Several experts noted that cyclists have historically been marginalised by policy and planning in the city, and little had been done to ensure their safety or ease of movement through the city.

The tiny numbers of existing cyclists, and the indications that many are forced by circumstance to forego public transport, suggests that new commuter cyclists will have to be drawn from different and more diverse groups in order to lead to a significant modal shift while acknowledging the existing challenges with distance. Several different strategies would need to be pursued.

Firstly, most trips in Gauteng are made to access work or to look for work, so integrating cycling with public transport routes to economic hubs is vital, especially for people who live in peripheral suburbs (Wray and Gotz, 2014). The greatest dissatisfaction for residents of Gauteng with their current mode of transport is high costs (Wray and Gotz, 2014). Cycling could therefore serve as a feeder mode to reach the nearest public transport or mini-bus pick-up point, particularly for people who wish to save on transport costs and who currently use private transport to reach Gautrain stations, bus stops or minibus pick-up points.

Secondly, a strategy focused on wealthier residents who are private-vehicle users could encourage people who already live close to work to cycle to reach their offices, or to conduct short trips for shopping or social visits instead of driving.

Lastly, an important constituency of potential cyclists are school children and students (though as cycling would primarily replace walking this would have limited impact on reducing carbon emissions). Existing NMT users feature a higher number of learners and students than other modes, with 10 per cent of NMT trips used to get to a place of study, compared to six per cent for other modes (Culwick, 2014). Currently, 51 per cent of school children walk to school, yet only two per cent ride bicycles. School transport is an important issue in the province as 37 per cent of households in GCRO's survey reported having one or more child in school. These could form 'lead' users in promoting commuter cycling.

One of the primary challenges that all of these strategies face is low levels of bicycle ownership and the inability to cycle – a 2010 survey conducted by the City of Johannesburg indicated that 91 per cent of respondents had no bicycle (though with the highest rate of bicycle ownership in relatively wealthy suburbs) (City of Johannesburg, 2014).

Days of extremely hot weather and summer thunderstorms, the city's undulating topography, as well as safety concerns from both other road users and criminals are often raised anecdotally as factors that would discourage cyclists. However, current NMT users do not list these as primary concerns.

The Development of Commuter Cycling Infrastructure and Policy

Despite the significant challenges such an initiative faces, the City has nonetheless chosen to champion cycling as part of a larger drive to improve transport. The history of this development demonstrates both key moments when leadership propelled efforts unexpectedly; others when key opportunities were missed; examples of innovative policies that were explored and failed; and long periods when efforts appeared to stagnate.

Most trace the beginning of these initiatives to work done at national level to reform transport systems across the country in 2009. This did not initially benefit an urban cycling agenda. A draft national policy released in 2009 framed the need for NMT infrastructure with reference to poverty and transport marginalisation, thereby making rural areas the priority for future interventions (South African Department of Transport, 2009). The Gauteng Provincial Government has nonetheless drawn on this policy – even if only in draft form – in designing its transport plans for the province, which is almost entirely urban (Culwick, 2014).

In 2010, the Planning Department of the City of Johannesburg itself decided to devise an NMT framework – a somewhat weaker piece of guidance than a fully-fledged policy – but this was drawn up by the Department of Development Planning, and failed to win over the Department of Transport.

More significant than these developments were changes to transport infrastructure and planning that were galvanised by the preparations for the Football World Cup in 2010. This gave momentum to plans to install a mass transit system with the government ultimately choosing to invest in a high speed rail link between the Oliver Tambo International Airport, the Rosebank and Sandton Central Business Districts (CBD) in Johannesburg and Pretoria, as well as a bus-rapid transit (BRT) system modelled on systems rolled out in Colombia and Brazil (Venter, 2013). While the core routes were completed before or shortly after the 2010 World Cup, both of these are still in the process of being expanded to achieve greater coverage.

When the plans and strategy for implementing were first drawn up, the City had the intention to include infrastructure in the stations and sensitivity in the planning of the network to integrate cyclists. But, according to officials who were involved at that period, time pressure prevented NMT concerns from being taken into great account or infrastructure for cycle storage included in the stations.

> *The big project for the time was the Rea Vaya Bus Rapid Transport. The focus on NMT was always a part of the vision for that project. Somewhere along the line, because we got so involved in taxi industry negotiations, the focus on NMT infrastructure got downplayed* (Interview, 28 May 2014).

Many respondents lamented that NMT had been an 'afterthought' to BRT and that opportunities to create cycling parking space or facilitate the integration of cycling with the BRT and Gautrain systems was not made possible. Currently, it is not possible to take a bicycle on BRT or Metrorail trains, and there is no safe cycle parking near their stations. Bicycles can only be brought on the Gautrain if they are in a bag, but even this loophole is not widely publicised and the guards will deny entry to anyone not complying with this rule (Hucthinson, 2014). In effect, it is impossible for cyclists to integrate their journeys with trips on public mass transit.[52] To this day, the City maintains that they cannot allow bikes on trains and buses because of safety concerns (bikes injuring other commuters) and the lack of space during peak hours.

Nonetheless, the measures taken during the World Cup catapulted Johannesburg's transport infrastructure and mainstreamed a range of 'green' concerns as well as a desire to present the City as one with 'modern' and 'global' infrastructure. It also created ring-fenced funding for transport and the requirement that new integrated transport plans include attention to cycling and walking facilities. Additionally, some of the cycle lanes coming to completion only now were also initiated in this period. The network of bicycle lanes in Soweto and Orlando have their origins in a grant offered by the German Bank for Reconstruction and Development (KWV) for projects that promoted 'green energy' (including energy savings or efficiency) that

52. It is unclear what the response from the minibus taxi industry would be, but the passenger would likely have to purchase multiple fares to make up for the space the bicycle occupied and responses would vary from operator to operator.

could be showcased at the World Cup. The Soweto-Orlando municipality won funding for their cycle routes but, due to the slow speed of local government implementation, these lanes were only completed in 2014, nearly four years later.

Between 2010 and 2014 there was little evidence of progress: infrastructure remained incomplete, safety hazards were acute, and a vital opportunity to achieve crucial integration with mass transit had been missed. If projects promoting the interests of pedestrians or cyclists got implemented they were generally ad hoc: 'NMT was an afterthought to Mass Transit. … Small NMT projects got done by individual ward councillors to get some political mileage as they were easily funded (Interview, June 8 2014).'

Advocates within and outside of local government also remarked upon what they perceived as strong 'hostility' to the idea that cycling could be promoted in Johannesburg in the face of the challenges outlined above. One informant, who had close involvement in transport policy work in local government, said: 'Part of the problem was the idea that "these are nice concepts from the rest of the world but they are not practical for Johannesburg" (Interview, June 8, 2014).' Another remarked that the Department of Transport responded to the cycle agenda as if it were 'fundamentally opposed to the identity of the City' (Interview, 8 June 2014). During the period before and shortly after the World Cup, at least one senior figure had made major strides with promoting 'eco-mobility'. But after this official left there was the sense that cycling lacked a champion within local government.

From around 2012, cycling advocacy groups began to lobby the city more forcefully to take measures to promote cycling. Many people who had been directly and indirectly involved in meetings between the city and these groups felt that plans and suggestions had been met with a high degree of negativity from local government; in return, civil society was also perceived to have unrealistic expectations about how long it takes for local government to implement decisions once they have been taken due, primarily, to the slow bureaucracy around tender processes. At times, dialogue between these civil society groups and local government threatened to break down.

Informants inside and outside of the city remarked on a shift in early 2014. Around this time an individual involved in cycling advocacy who had good lines of communication with local government put forward a proposal for an event that would promote commuter cycling under the banner of commemorating the death of former president Nelson Mandela. The 'Freedom Ride' would symbolically link the heart of the city with the formerly marginalised, outlying suburb of Soweto (Freedom Ride, 2014).

In 2014, several other plans that had been under development began to bear fruit. Cycle lanes linking the city's main universities began to be constructed and cycle lanes in Soweto reached completion. Initially, these lanes will be shared with other road users; over time, they will be upgraded and dedicated for cyclists' use only, and some will have physical barriers that separate them from motor vehicle lanes.[53] These have been accompanied by strong statements of support for the commuter cycling agenda by the office of the mayor, such as repeated statements that: 'Over time, we will eliminate the need for private vehicles as the city progressively moves towards an effective public transport system, cycling lanes and pedestrian walkways (Güles, 2013) see also (Samuels, 2014; Tau and Bloomberg, 2014).'

The city met with the universities to discuss the promotion of cycling on campuses and met monthly with a range of civil society actors at the Jozi Cycle Forum to discuss plans, including a way of finding a system and the sponsorship of a 'Bicycle Empowerment Network' centre in Soweto which would offer affordable bicycle repair. A second Freedom Ride was held in June 2014. Plans for 2015 are even more ambitious.[54]

Several policies have also consolidated around the push for NMT: e.g. 'Complete Streets' and the 'Corridors of Freedom'.[55] The challenges to the adoption of cycling appear to be acknowledged. The City of Johannesburg acknowledges the challenge that low-density urban sprawl poses for the cycling agenda by limiting its investment to connecting cyclists with destinations 2.5km to 6km from their homes or workplace (City of Johannesburg, 2014). On paper, the city's approach to cycling encompasses infrastructure as well as behaviour change (safety training and promotion) and access to bicycles. In practice, the provision of infrastructure is in a more advanced state than the other 'pillars' (see Box 1) and, anecdotally, cycle lanes alone do not appear to be translating into greater numbers of trips by bicycle.

Efforts to promote 'bicycle friendliness' therefore sit at an interesting juncture. Considerable progress has been made since the days when cycling

53. These will link the University of Johannesburg and the University of the Witwatersrand, and lanes will also be laid down in Orange Farm and Ivory Park. These join 5.5 km of lanes almost complete in Orlando, Soweto.

54. Two large cycling promotion events will take place: a Freedom Ride, preceded by a week's worth of events and a large NMT conference, and an 'Eco-Mobility', co-sponsored by ICLEI, which will shut down parts of the busy Sandton precinct to cars. Fifty km of 'Complete Streets' will be laid down in Hillbrow, Sophiatown and Langlaagte, and dedicated cycling infrastructure, including a cycle bridge over a busy highway, are scheduled to be constructed in 2015 (City of Johannesburg, 2014).

55. Corridors of Freedom policy aims to direct economic development towards major transport arteries and Complete Streets aims to redesign streets to accommodate pedestrians, cyclists, the disabled and a range of users other than cars.

was 'sidelined' and the city attitude was characterised as 'hostile'. But the range of initiatives currently being pursued has not had time to bear fruit. Cycle lanes in particular raise a specific behaviour challenge: unlike an environmentally destructive practice that can be regulated away, these require a positive and spontaneous response from citizens.

Box 1: The City's Cycling Pillars:

Construction of cycle-friendly infrastructure
Dedicated cycle lanes
Bike storage and bike parking points
Bikes on public transport

Making cycles more accessible
Donation of bikes to learners
Enabling the private sector to produce bicycles that are more robust
Ensuring that bike repair is affordable

LESSONS FOR CITIES: WHAT DOES THIS TELL US ABOUT 'GREEN' CITY-LEVEL CHANGE IN SOUTH AFRICA?

International Networks and 'Policy Entrepreneurs'

The international literature on urban climate change governance puts a lot of emphasis on the role of both 'policy entrepreneurs' and international networks in promoting action. To a certain extent these two phenomena dovetail. Bulkeley argues that: 'Networks have provided the resources and political space within which policy entrepreneurs can operate with some degree of protection from "politics as usual" and that they offer enterprising leaders "soft" rewards for pioneering actions and trigger events, such as the hosting of global conferences or sporting events (Bulkeley, 2010: 244).'

In Johannesburg, Mayor Parks Tau's support for cycling has been demonstrated most prominently through his involvement in the C40 network – and Johannesburg's hosting of the C40 City Mayors Summit in February 2014 – and his upcoming plans for the city to host a high-profile 'eco-mobility' festival in Sandton in 2015 in partnership with ICLEI. Several informants in civil society attributed the shift in city officials' attitudes to the public support the mayor has given to this agenda, including 'getting on a bike' to join the Freedom Ride (Interviews, 19 May 2014, 25 May 2014, 27 November 2014). However, several other respondents were cautious about endorsing this idea. Firstly, because there was diverging opinion about whether the mayor's support had involved tangible expenditure of political capital, or if it was more a case of an opportunistic recognition of the political mileage to be gained from associating his tenure with infrastructure which, while initiated many years ago, is only bearing fruit now. It is also clear from the narrative described above that the city's Department of Transport has been the primary power holder on the issue of cycling. This department comes with its own political head in the form of the MMC for Transport. Department of Transport staff emerge as having a real acceleration or chilling effect on policy implementation.

People with direct experience of participating in international networks were surprisingly mixed in their assessment of these, too. Several informants commented that they felt these forums were often used to push private sector interests, or the political careers of particular high-profile participants without a clear benefit for participating cities. According to one local government staff member, one set of meetings had been used to push the sale of electric cars, raising questions about whose agenda was being served: 'To what extent are the discussions on sustainability and environment in these forums driven by economic interest over and above what is good for the planet? (Interview, 28 May 2014).' Another informant described them as 'a cheap and easy to way to tap into a "world class narrative" in which cycle lanes are a "buzzword" (Interview, 22 September 2014)'. At the same time, networks were seen to have a following in disseminating ideas and practices across the world. 'These forums are useful for the sharing, replication and modification of ideas and in order for cities to have a sense of clarity about their own areas of focus (Interview, 28 May 2014).'

More fundamentally, officials within the city downplayed both the role of international networks and the support of the mayor and attributed their own watershed moment with the cycling agenda to the Freedom Ride. The fact that it was seen as a 'success' and generated good publicity for the administration amongst local citizens had warmed officials to cycling, which they had previously worried was a niche interest.

Civil Society and Deliberative Democracy

Action by the city has been tied to pressure and incentives from external organisations in several respects, and much of it was the fruit of projects initiated many years previously. The first phase of cycle lanes connects the CBD with the University of the Witwatersrand and the University of Johannesburg, and this location has its origin in a proposal brought forward by the universities themselves to improve student transport (not specifically cycle lanes). The cycle lanes in Orlando, as mentioned above, have their origins in German development financing.

The advocacy group JUCA has also played an important technical role. JUCA, which 'represents bike commuters and aims to promote a bike-friendly Jo'burg' researched and designed a set of possible cycle routes for the city which would 'link up residential and commercial centres, and which can be established comparatively easily with a limited and specific infrastructure intervention by the City' (JUCA, 2014). They have also proposed a way-finding system to accompany this – a proposal the City has accepted and assigned funding towards. Additionally, they have been consulted for technical input into the design of the cycle lanes being rolled out and have provided feedback on their experience of the lanes as cyclists (Iqani, 2014). A range of other groups also contributes to promoting commuter cycling, for example, through organising recreational rides that introduce people to urban cycling through forming partnerships with the private sector to promote cycling amongst employees, and charity activities that distribute bicycles to high school learners (see Box 2 for more).

Civil society groups were also instrumental in setting up a forum for discussion between the city and other stakeholders: the Jozi Cycle Forum. The aim of the forum is to 'promote cycling in all its forms and share information between, support, coordinate and track the various cycling projects currently underway in the City of JHB' (Iqani, 2014). This forum has been important in developing the partnerships needed for the Freedom Ride, the way-finding system, and other developments.

While the relationship between the city and NGO groups has been – and may continue to be – volatile, an independently facilitated forum appears to have laid the groundwork for important interactions.[56] 'For me it has been one of the most constructive partnerships with civil society. They don't ask for money and stipends. They come into town to meet us ... ' explained one

56. One cycling advocate described how having the city facilitate the meeting had almost caused the discussion to break down: 'we used to have a logjam with the city ... it didn't work having the city as facilitator. They were convening meetings but also having to defend themselves (Interview, 19 May 2014).

Box 2: Cycling Advocacy Groups

The promotion and future of commuter cycling is not only related to the City's policy and infrastructure, but also concerns a range of other lobbies and groups which seek to change attitudes to cycling, secure investment in cycling infrastructure from elsewhere, or simply to establish or expand an everyday cycling culture that they enjoy being a part of. These groups include a local version of Critical Mass, which also organises rides in Pretoria (http://jhb.criticalmass.co.za/). The Johannesburg version of this global phenomenon is less political than many international chapters and is associated with being 'more of a party ride', with a motorbike cavalcade protecting the ride against incidents with vehicles. The Monthly Cycles is a small group ride that meets once a month on full moon and was formed in counter-culture to what its founders perceived as a macho and competitive element present in Critical Mass. It is loosely organised and geared towards exploration of particular neighbourhoods. Fixin' Diaries is one of a few fixed-geared orientated groups operating in Soweto and the CBD. Its members are street fashion and fixed gear enthusiasts, linked to commercial outfits with an inner-city aesthetic (http://www.fixindiaries.co.za). Networks that organise even more informal group rides are orientated around a growing number of boutique bicycle stores. Operating in a rather different mode is 'Hash Tag Decongest Sandton', a campaign linked to the electric bicycle business 'Cycology' (Brodie, 2014). The campaign is orientated towards the promotion of electric bicycles in the upmarket business district of Sandton. Its business case is framed around the loss of profit caused by late employee arrivals and low employee well-being due to traffic congestion.

city official. 'Sometimes they drive me up the wall with all the debates … but the commitment, the volunteerism is great. I think that's what makes Johannesburg a dynamic city (Interview, 30 October 2014).' An advocate emphasised that they saw their role as representing groups who would not otherwise engage: 'It's important to build a constituency of people who will raise their voice. People have been commuting by riding bikes for years, but they remain very marginalised (Interview, 28 May 2014).' A former government employee remarked that lobby groups had been important in creating accountability: 'It's the ability to ask the right questions, to make sure the projects committed to have been implemented, and that, as consultant and professional teams are brought on board, the voices of community groups are not lost (Interview, 28 May 2014).'

Equity Concerns

The City of Johannesburg has explicitly framed cycling as an intervention that achieves several objectives, many of which are not environmental. A promotional presentation in current use by the city identifies the following four points as core elements of the 'business case' for greater investment in NMT:

- Reduces congestion[57]
- Reduces CO_2 emissions, improves air quality and reduces noise in our neighbourhoods
- Improves fitness and health of our communities
- More people on our streets walking and cycling breaks down barriers and improves active citizenry

The shift to promoting a greater modal share for cycling is presented as resolving citywide problems and social cohesion. So far, its concrete manifestation has been through the construction of bicycle lanes. Lanes, while taking the safety of cyclists seriously, do not solve the problem of distance raised above in relation to the city's unequal development. In Europe, four to six kilometres is considered an acceptable cycling distance and most people will not commute longer than this. In South Africa, average commutes are at least five to 10 kilometres,[58] with people coming from outlying townships travelling far further. The distance between Soweto and the old Johannesburg CBD is around 25km – the distance from Soweto to the northern CBD, Sandton, is around 40km. One policy expert commented: 'If your trip is 25km, a bike lane isn't going to make that trip any shorter, or to allow you to pick up a child or go shopping. In a place like Jo'burg, the issue is really about distance and trip chaining. So a bike lane won't really make a difference (Interview, 22 September 2014).'

The city acknowledges this reality and aims to first target those people

57. Of all of these, congestion is perhaps the most directly motivational for the City, lobby groups and business. The province's population has doubled in the last 20 years and will double again in the next 25, and the working population itself will increase to 1.75 times its current size (Second Non-motorised Transport Conference, 2014). A recent travel survey by the City of Johannesburg noted that the proportion of people using private vehicles rose from 53 per cent in 2010 to 57 per cent in 2014. The average trip time is 51 minutes (Minutes of Jozi Cycle Forum Meeting).

58. Statistics provided by a cycling policy expert interviewed for this paper.

who are travelling shorter distances, and this can be effective at linking people for discreet trips (e.g. to school, the shops, to work if nearby). Cycling infrastructure has also been located both in the multi-ethnic and mixed class CBD, and in predominantly low-income suburbs like Soweto (though in the more middle-class area of Orlando), and the plans for infrastructure leading out of Alexandra are in the final stages of completion. However, it is clear that cyclists, particularly in historically marginalised outlying suburbs, will not be integrated in a citywide transport network unless they can combine journey by bicycle with trips on public transport.

Another crucial determinant of whether cycle lanes are used will be bicycle access, which is currently heavily class based. A 2010 survey by the City of Johannesburg revealed that 91 per cent of the city's inhabitants do not own a bicycle. Research done by the city indicates that a sturdy bike would cost a household a minimum of R2,150 (City of Johannesburg, 2014). In a city where many households live near the poverty line, this is unaffordable. This currently presents perhaps the most difficult problem facing the cycling agenda. While the city was initially enthusiastic about using affordable bike-share schemes to get around this problem, extensive feasibility studies have deemed this to be an unrealistic option for the city, highlighting concerns about affordability, security and safety, as well as the persistent problem of low-density development and long commute distances (City of Johannesburg, 2014). An existing government initiative that donates bicycles to learners, the Shova Kalula programme, has had poor results, having woefully failed to meet targets for distribution and with a number of the bicycles being sold on or falling into disrepair when those who receive them cannot afford to fix them or have no access to bicycle mechanics.

The risks of neglecting these aspects of cycling promotion are acute, as policy failure may result in a backlash. One informant in local government expressed his concerns as such:

> *The other worry for me that might cause problem[s] is the cost of implementing the cycle lanes. In Diepsloot [a poor township on the outskirts of the city] there is no clinic. With 20 million you can build a clinic. Or provide toilets. Or water. The cycle lanes don't come cheap. One day somebody is going to say – maybe the Economic Freedom Fighters – why are all these bicycle lanes here and no one is using them?* (Interview, 30 October 2014).

CONCLUSIONS

The City of Johannesburg's progress with implementing a cycling-friendly agenda has shown great momentum, which continues to build. But it is not yet fully addressing major blockages, such as poor access to bicycles and lack of integration with public transport, which will enable potential users to make use of its investment in infrastructure.

At a more general level, the development of commuter cycling infrastructure and policy in Johannesburg demonstrates some of the internal politicking in local government that allows such schemes to be championed and implemented, and which may also slow or stall progress. These processes do not happen independently of outside actors, which include a range of lobby groups that can serve to increase accountability or to divert policy towards parochial concerns.

Lessons can be drawn for other cities and other policy issues that require galvanizing action from a range of actors and encouraging citizen behaviour change under an umbrella of supportive policy led by local governments.

Recommendations:

- Current cyclists, while not numerous, are a constituency which has been marginalised in policy and debate. More hard data should be collected about when, how often, and under what conditions people currently cycle. Data should also be collected and monitored about road accidents involving cyclists.
- With cycle lanes underway and nearing completion for many key routes, the primary obstacles to greater uptake of cycling lie in bicycle access and integration into a functioning mass transit system.
- Making bicycles and their repair affordable for low-income citizens should be a major priority of further work. As bicycle-share schemes have been found not to be feasible, this is likely to rest on models that rely on the state (or its NGO partners) providing bicycles at little or no cost, as in the model put forward by Qhubeka.[59] Solutions could also be sought through partnership with the private sector, such as through a 'Ride to Work' as in the UK, though this only reaches people in formal employment.
- The integration of bicycles into other forms of public transport, such as the trains and buses, is crucial to linking cyclists in far-flung suburbs

59. Qhubeka is a not-for-profit organisation that donates hardy bicycles to its target groups (including school children and health care workers) in exchange for their contributions through community work or as part of a contract involving, e.g. school attendance. http://qhubeka.org/2013/?page_id=30

with destinations in the city centre. While space constraints may prevent this in peak hours, a system could be devised to allow bicycles on buses, trains and the Gautrain during off-peak hours initially. This should be accompanied with safe bicycle parking at stations.

- Given the large challenge posed to eco-mobility solutions such as cycling by the vast distances created by South Africa's incredibly low-density urban development, the commuter cycling cause must also be linked to demands to improve the public mass transit system and also for policies that encourage densification of existing settlements.
- The partnership between the state, private sector, universities and lobby groups feeds into accountability and the momentum for policy implementation, as well as improving the content of policies by providing 'user' input. Two initiatives appear to have galvanized and smoothed this dynamic: an event, in the form of the Freedom Ride, for different actors to lobby around, and a neutral forum to air grievances and coordinate action in the form of the Jozi Cycle Forum. These could be replicated in other cities to strengthen momentum towards promoting cycling.
- International networks, on the other hand, have had much more muted value. While the value they have added in disseminating ideas and providing incentives for cities to excel as 'pioneers' in environmental innovation should be further examined, there is also an argument for directing investment (of time and personnel) towards extending partnerships at the local level instead. For example, with groups concerned with the safety of (child) pedestrians en route to school, or small businesses that could benefit from foot (and cycle) traffic.

References

Aylett, A. 2010. 'Conflict, Collaboration and Climate Change: Participatory Democracy and Urban Environmental Struggles in Durban, South Africa', *International Journal of Urban and Regional Research*, 34: 478–495. DOI: 10.1111/j.1468-2427.2010.00964.x.

Bulkeley, H. 2010. 'Cities and the Governing of Climate Change', *Annual Review of Environment and Resources*, 35, 229–253. DOI: 10.1146/annurev-environ-072809-101747.

Bulkeley, H. and Schroeder, H. 2008. 'Governing Climate Change Post-2012: The Role of Global Cities – London', Tyndall Centre Working Paper 123. Oxford: Tyndall Centre. Available at: www.tyndall.ac.uk/content/governing-climate-change-post-2012-role-global-cities-london [Accessed 10 May 2015].

City of Johannesburg. 2014. 'Making Joburg a cycle friendly City: a call for partnership', Powerpoint presentation used by the City of Johannesburg throughout 2014.

Department of Environmental Affairs (DEA). 2013. *GHG Inventory for South Africa 2000–2010*, August 2013, research report commissioned by the DEA and published online. Available at: https://www.environment.gov.za/sites/default/files/docs/greenhousegas_invetorysouthafrica.pdf [Accessed 9 January 2015].

Freedom Ride. 2014. *Freedom Ride – Jozi 2014*. Available at: www.freedomride.co.za/joburg/ [Accessed 10 May 2014].

Gore, C. D. and Robinson, P. J. 2009. 'Local Government Responses to Climate Change: Our Last, Best Hope?', *Changing climates in North American Politics: Institutions, Policymaking and Multilevel Governance*. Ed. Henrik Selin and Stacy VanDeveer. Cambridge: MIT Press. 137–158.

Green, E. 2013. 'Can Mayors save the World?' *The Atlantic*.

Greenfield, A. 2014. 'Buses are the future of urban transport. No, really'. *The Guardian*.

Güles, N. 2013. 'Newsmaker – Parks Tau: My vision for Joburg'. *City Press*.

Hutchinson, J. L. 2014. 'How to smuggle a bicycle onto the Gautrain'. Johannesburg: Urban Cyclists Association.

Iqani, M. 2014. 'Cycling Lanes are coming to Johannesburg'. Johannesburg: Urban Cyclists Association.

Iqani, M. 2014. 'Feedback on Braamfontein Cycle Lanes'. Johannesburg: Urban Cyclists Association.

Jacobs, E. 2013. 'Why mayors have more chance of saving the world than global leaders do'. TED Blog.

JUCA, n.d. 'About JUCA'. Johannesburg: Urban Cyclists Association.

Kennan, H. 2014. 'Cities Step Up in Fight Against Climate Change', *Energy Innovation*. Available at: http://energyinnovation.org/2014/10/cities-step-up-in-fight-against-climate-change/ [Accessed 10 May 2014].

Lutsey, N. P. and Sperling, D. 2008. 'America's Bottom-Up Climate Change Mitigation Policy'. Institute of Transportation Studies. UC Davis: Institute of Transportation Studies (UCD).

Mukheibir, P. and Ziervogel, G. 2007. 'Developing a Municipal Adaptation Plan (MAP) for climate change: the city of Cape Town', *Environment and Urbanization*, 19: 143–158.

DOI: 10.1177/0956247807076912.

Municipal IQ. 2014. '2014 protests may reach new peak', 4 September.

Phakathi, B. 2014. 'Zille slams politicisation of Western Cape projects', *Business Day Live*, 8 August.

Roberts, D. and O'Donoghue, S. 2013. 'Urban environmental challenges and climate change action in Durban, South Africa', *Environment and Urbanization*, 25: 299–319. DOI: 10.1177/0956247813500904.

Samuels, S. 2014. 'Joburg set to become more bicycle friendly', *Midrand Report*.

Seijas, A. 2014. 'Cities for People', *Americas Quarterly*, Winter 2014.

South African Department of Transport. 2009. 'Draft National Non-Motorised Policy'.

Tau, P. and Bloomberg, M. 2014. 'Green cities can help breathe new life into nations' growth', *Business Day Live*. Available at: www.bdlive.co.za/opinion/2014/10/13/green-cities-can-help-breath-new-life-into-nations-growth [Accessed 10 May 2014].

Taylor, A., Cartwright, A., Sutherland, C. 2014. 'Institutional Pathways for Local Climate Adaptation: A Comparison of Three South African Municipalities'. Agence Francais de Developement.

The Economist. '2013 Cycling in New York: End of a lovely ride?' 3 December. Available at: http://www.economist.com/blogs/democracyinamerica/2013/12/cycling-new-york [Accessed 9 January 2015].

UN Habitat, 2011. 'Climate Change Mitigation Responses in Urban Areas'. Geneva: UN Habitat.

Venter, C. 2011. 'Transport expenditure and affordability: The cost of being mobile'. Development Bank of South Africa, 28, 121–140. DOI: 10.1080/0376835X.2011.545174.

Venter, C. J. 2013. 'The lurch towards formalisation: lessons from the implementation of BRT in Johannesburg, South Africa', *Research in Transportation Economics*, Vol. 39, 1: 114–120. DOI: 10.1016/j.retrec.2012.06.003.

Walker, P. 2013. 'Boris Johnson vows to continue London's "cycling revolution"', *The Guardian*, 9 December 2013. Available at: http://www.theguardian.com/uk-news/bike-blog/2013/dec/09/boris-johnson-london-cycling-revolution [Accessed 9 January 2015].

Wray, C. and Gotz, G. 2014. 'Mobility in the Gauteng City-Region'. Johannesburg: Gauteng City-Region Observatory.

ANNEXURE: PEOPLE INTERVIEWED AND ORGANISATIONAL AFFILIATION

1. Interview with Njogu Morgan, JUCA, 28 May 2014.
2. Interview with Rehana Moosajee, former MEC for Transport, Gauteng, 28 May 2014.
3. Interview with Christina Culwick, GCRO focal point for NMT, 2 April 2014.
4. Interview with Geoff Bickford, SACN, 3 April 2014.
5. Interview with Thomas Coggin, Urban Joburg, 31 July 2014.
6. Interview with Carolyn Frick, formerly with the Department of Planning, City of Johannesburg and author of the NMT guidelines, June 8 2014.

7. Interview Gail Jenning, independent consultant specialising in NMT, 22 September 2014.
8. Interview with Marcela Gerrero Casa, Director of Open Streets Cape Town, 7 September 2014.
9. Interview with Melvin Neale, co-organiser of Critical Mass, 25 May 2014.
10. Interview with Michael Flanagan: Masters Student at WITS focusing on Eco-mobility, 31 July 2014.
11. Interview with David Du Preez, JUCA, 27 November 2014.
12. Interview with Mehita Iqani, Monthly Cycles, 27 November 2014.
13. Interview with Hastings Chikoko, C40, 15 April 2014.
14. Interview with Lisa Seftel, Executive Director, Department of Transport, City of Johannesburg, 30 October 2014.
15. Interview with Crispin Olver, LinkD and Chair of Jozi Cycle Forum, 19 May 2014.
16. Interview with Simphiwe Ntuli, City of Johannesburg, 30 October 2014.
17. Interview with Esther Lethlaka, City of Johannesburg, 30 October 2014.
18. Interview with Daisy Dwango, City of Johannesburg, 30 October 2014.

CHAPTER 7

The Energy and Water Nexus: The Case for an Integrated Approach for the Green Economy in South Africa

Manisha Gulati

INTRODUCTION

The fundamental requirement of any economy is that it delivers food, water and energy security for all. In a green economy this will be provided by renewable energy; green buildings, including green retrofits for energy and water efficiency; clean transportation utilising alternative fuels, hybrid and electric vehicles (see Mytelka's Chapter 5); water management, including water demand management and conservation, water reclamation, purification and recycling such as industrial and domestic effluent; waste management (see Pilusa and Mazenda's Chapter 8), brownfield remediation and sustainable packaging; and land management, including organic agriculture, habitat conservation and restoration, urban forestry and parks, reforestation and afforestation, and soil stabilisation (PwC, 2011). From the foregoing, it is clear that water and energy are key components of the green economy.

At a simplistic level, the relationship between water and energy pertains to energy intensity in the water sector or the amount of energy needed for extracting, treating, transporting and distributing water, disposing of waste water, and water intensity in the energy sector or the amount of water needed for harnessing, extracting, producing and transmitting energy. Decisions about one can exacerbate scarcity, or enable better management of the other from the perspective of conservation, consumption efficiency, and reuse and recycling (of water). The energy-water relationship, therefore, goes beyond simple footprint calculations and involves an understanding of the interdependencies and complications of water and energy.

This relationship or nexus of water and energy is one of the key entry points into the green economy debate and solutions. The philosophy of the green economy requires increased resource use efficiency, improving energy and water security, securing sustainable access to water and energy, and maximising the social amenity of energy and water. A nexus approach will support the transition to a green economy by creating opportunities for addressing synergies and trade-offs across these two sectors and enhancing policy coherence. Appreciation of this nexus will also highlight the set of green investment opportunities that are starting to emerge.

The objective of this chapter is therefore to highlight how an integrated approach to water and energy is one of the critical ways to achieving the green economy in South Africa. In doing so, it will specifically focus, firstly, on the gaps in policy and planning from a nexus perspective and, secondly, on how addressing these gaps could be pivotal to achieving a green economy.

The chapter is organised as follows: the first section explores the relevance of the energy and water nexus (EWN) to a green economy. It looks at the various elements and concepts of the green economy to explore the importance and role of the EWN in delivering this economy. Addressing any debates over the concept or application of the green economy approach or debating the merits and demerits of this concept is beyond the scope of this chapter. The second section provides an in-depth understanding of the interdependencies between energy and water. The third section looks at the energy and water-related challenges in the context of South Africa. The final section provides recommendations on how integrated energy and water planning could help address these challenges, thereby helping achieve the green economy objectives.

ENERGY AND WATER NEXUS AND THE LINK TO A GREEN ECONOMY

The Concept of the Green Economy

A green economy basically refers to a low-carbon economy with an efficient use of natural resources as well as traditional inputs such as labour and capital to produce well-being for the population. The philosophy underpinning the concept involves:

1. investing in, protecting and building the natural resource management and ecological restoration, particularly in the natural systems on which poor and indigenous communities depend for their livelihoods;
2. allocating environmental benefits and costs fairly for the achievement of a just and equitable society;
3. green economic services and industries incorporating efficiency gains in the quantity and quality of production that provides decent work, new employment prospects and affordable sustainable consumption alternatives; and
4. resource-efficient and low carbon economic development (Gulati, 2014b).

Relevance of Energy and Water Nexus to Green Economy

The term EWN is used to describe the interdependent, synergistic and mutually reinforcing relationship of energy and water. The production cycle for fuels and energy requires water at various stages: fuel extraction (mining and refining; oil, gas, uranium and coal processing; and coal and gas liquefaction and gasification), and cultivation of crops for energy production and generation (see Figure 1). Energy extraction and production also have an impact on water availability and quality. Similarly, energy is required at all stages of the water-use cycle (see Figure 2). Large amounts of energy are required to pump, treat and distribute water for urban, industrial, and agricultural use, human use, and to deal with the resulting waste. In fact, relative to its value, water is heavy and, in energy terms, it is expensive to pump water over long distances as well as to lift it (UNEP, 2011). Energy is used by agriculture for accessing water resources, especially groundwater through pumping, and by households and industry for heating and cooling water (Gulati, 2014a). It is also used to purify and soften water for household use (Gulati, 2014a).

Figure 1: Water-use cycle

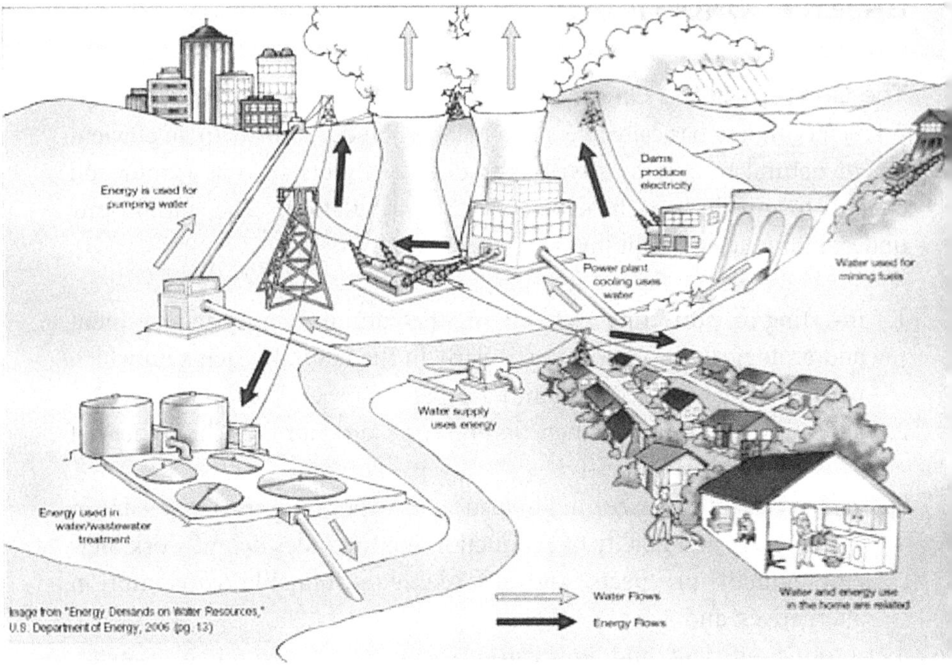

Source: US Department of Energy (DoE), 2006

Figure 2: Water-use cycle

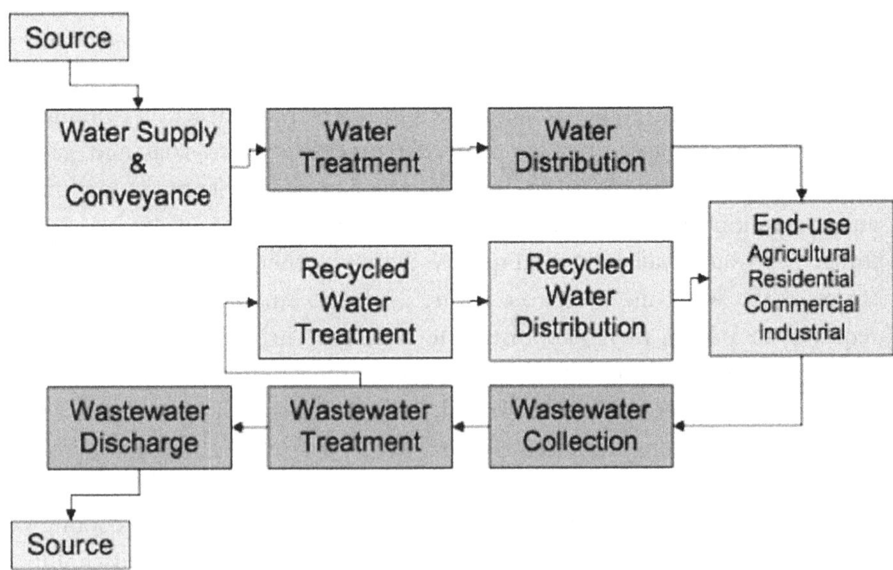

Source: California Energy Commission (2005)

Table 1: Role of water and energy in poverty alleviation and human development as represented through the Millennium Development Goals

MDG Goal	Direct contribution of energy for achieving the MDG	Direct contribution of water for achieving the MDG
MDG 1: Eradicating extreme poverty and hunger	• Increased production, business development and income saving	• Factor of production in economic activity • Direct input into irrigation for grain production, for rearing livestock, tree crops, and subsistence agriculture
MDG 2: Universal primary education	• Improved educational environment through water, sanitation, lighting and commuting to school	• Improved school attendance rates on account of access and provision of safe, secure and clean water services
MDG 3: Promoting gender equality and empowering women	• Less time spent by women in collecting firewood and performing manual farm labour	• Opportunities for productive activities for women and girls and improved social and economic capital for women in terms of leadership, earnings and networking opportunities with improved water services • Lower risk of sexual assault of women and girls when gathering water and searching for privacy with location of water and sanitation facilities closer to homes
MDG 4 and 5: Reducing child mortality and improving maternal health	• Support access to health services, improving clean water availability and reducing water-borne diseases	• Lower maternal mortality risks due to improved health and reduced labour burdens from water portage and lower morbidity and mortality factor for children due to improved quantities and quality of domestic water and sanitation

MDG Goal	Direct contribution of energy for achieving the MDG	Direct contribution of water for achieving the MDG
MDG 6: Combating HIV and AIDS, malaria and other diseases	• Enable refrigeration of medicines, a distribution system for medicines, and access to health education through ICT	• Better water management for reduced mosquito habitats and malaria incidence as well as incidence of a range of diseases caused by poor water management
MDG 7: Ensuring environmental sustainability	• Enable mechanical power in agriculture to reduce demand for land expansion; reduce dependence on biomass; reduce environmental impact of energy sector	• Improved water management, including pollution control and sustainable levels of abstraction helping maintain ecosystems integrity

Source: UNECA, 2013; Paul, 2003; and Human Development Report, 2006

Clearly, energy is a water issue and water is an energy issue. Increase in demand for one will increase the demand for the other while shortage of the one can constrain the availability of the other. This is problematic given that both water and energy are crucial for human development, poverty alleviation and economic development (see Table 1).

From a planning and management perspective, the interdependencies between energy and water, coupled with increasing demands for energy, the need to lower carbon emissions, and the reduced availability of fresh water supplies pose considerable challenges to ensuring the sustainability of both resources. For example, policies to reduce carbon emissions such as carbon capture and sequestration could increase water use and policies for promotion of biofuels for transportation could increase competition for water resources. Irrigated first-generation soy- and corn-based biofuels can consume thousands of times more water than traditional oil drilling (Glassman, *et al.* 2011). The supposed low water intensity of second- and third-generation biofuels is yet to be proven. Similarly, harnessing new, distant or poor quality water supplies and stringent standards for water and wastewater treatment could increase energy requirements.

At the same time, measures to support or enhance the conservation, consumption efficiency, and reuse and recycling of water or another resource could have unintended consequences for the other. For example, increasing energy supplies through certain types of incentives, or the lack thereof, or subsidising energy supplies could have unintended negative impacts on the national or regional availability of fresh water or water quality unless these policies are closely evaluated for both energy and water impacts. Similarly, efficient energy production facilities lead to more energy produced per unit of water used.

Figure 3: Relevance of the Energy and Water Nexus to the Green Economy

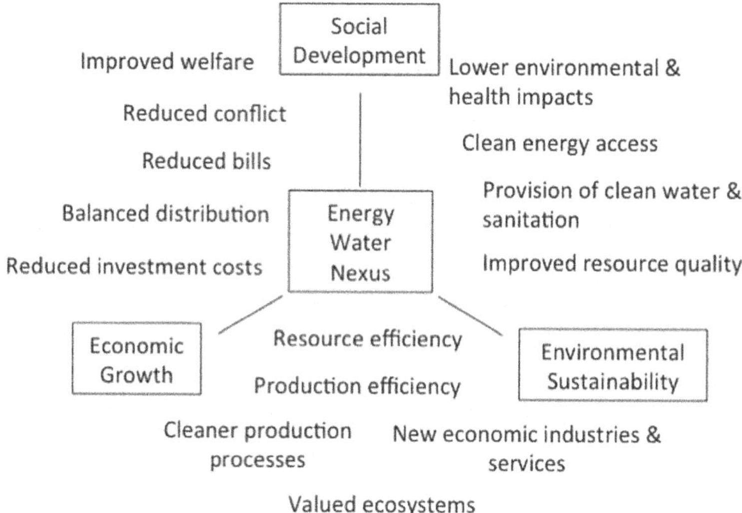

Source: Author

The EWN is therefore an important entry point into the green economy debate. The framework in Figure 4 conceptualises the importance of the EWN from the perspective of the green economy. The EWN will provide a systems perspective that is key to supporting the transition to a green economy through a better understanding of natural resource requirements for water supply and energy production; increasing social justice; decreasing poverty; increasing efficiency in the use of these resources; identifying synergies and addressing trade-offs across these two sectors; optimal allocation of these resources between sectors, geographical areas and population groups; enhancing policy coherence; optimising investments; and delivering high-quality energy and water-related services with less use of resources to support social and economic development in an eco-efficient and inclusive manner.

In doing so, the EWN will improve energy and water security; secure sustainable provision of water and energy; link growth, equity, poverty and sustainability; ensure that future resource needs will be met; improve economic efficiency across multiple sectors that use these resources; and help build resilience. Appreciation of this nexus will also highlight the set of green investment opportunities that are starting to emerge. Some such opportunities could be low water footprint energy technologies and low energy footprint water technologies.

UNDERSTANDING THE ENERGY AND WATER NEXUS

The Nexus From an Energy Perspective

From an energy perspective, the link between energy and water refers to the water requirements of energy technologies as well as the impact of energy extraction and production on water quality. Energy production requires water at the stages of fuel extraction, transportation, processing and refining, as well as at the energy-generation facility level. This water requirement is typically defined by way of water withdrawal and water consumption. Water withdrawal refers to the amount of water that is removed from the ground or diverted from a water source for use, but does not indicate the amount that is returned to the source after use. Water consumption refers to the amount of water that evaporates, transpires, is incorporated into products or crops, or is otherwise removed from the immediate water environment (Macknick, *et al.* 2011) but not returned to the source.

Water requirements for fuel extraction, transportation, processing and refining, and the water-use implications associated with the land required for infrastructure construction for energy generation (Pegasys, 2011) (Table 2) suggest that upstream activities related to coal require the maximum amount of water. Coal mining requires significant amounts of water for beneficiation (coal washing), equipment cooling and lubrication, dust suppression, site operations (potable water) and post-mining replanting of vegetation (Eskom, 2011 and US DoE, 2006). In practice, this water consumption is included under the industrial or mining sector and is not reflected under the water intensity of the electricity-production technology.

Withdrawals for fuel refining and transport are relatively small compared to those for thermoelectric cooling, but are still significant (American Geophysical Union, 2012). Oil refineries consume about 880 MGD (million

Table 2: Water requirements for upstream activities related to coal-, gas- and nuclear-based electricity production

Lifecycle stage	Withdrawal (gal*/MWh)	Consumption (gal*/MWh)
Coal		
Mining/processing	58	16
Transport (slurry pipeline)	473	170
Plant construction	7	N/A
Total	538	186
Gas		
Extraction/purification	44	15
Transportation/storage	14	8
Environmental control	235	N/A
Total	323	23
Nuclear		
Mining/processing	66	19
Plant construction	8	3
Spent-fuel disposal	5	N/A
Total	**79**	**40**

*One gallon = 3.78 litres

Source: Wilson, *et al.* (2012)

gallons per day) of water (about 1 gallon of water for each gallon of oil refined), and natural gas refining and pipeline transport consume about 400 MGD (American Geophysical Union, 2012).

At the energy production stage, energy technologies differ on water withdrawal and water consumption requirements. These requirements depend on factors such as fuel type, quality of raw water, quality of fuel and processing needs (American Geophysical Union 2012) and vary substantially even within technology categories. Therefore, these requirements are represented through a range rather than a specific single number. This also makes the comparison of technologies difficult. Therefore, electricity-generation technologies are often compared on the basis of water requirements for each unit of electricity generated (Figure 4), while transportation fuels are compared by water usage per unit of energy produced (Table 3).

Figure 4: Water use for electricity generation by cooling technology

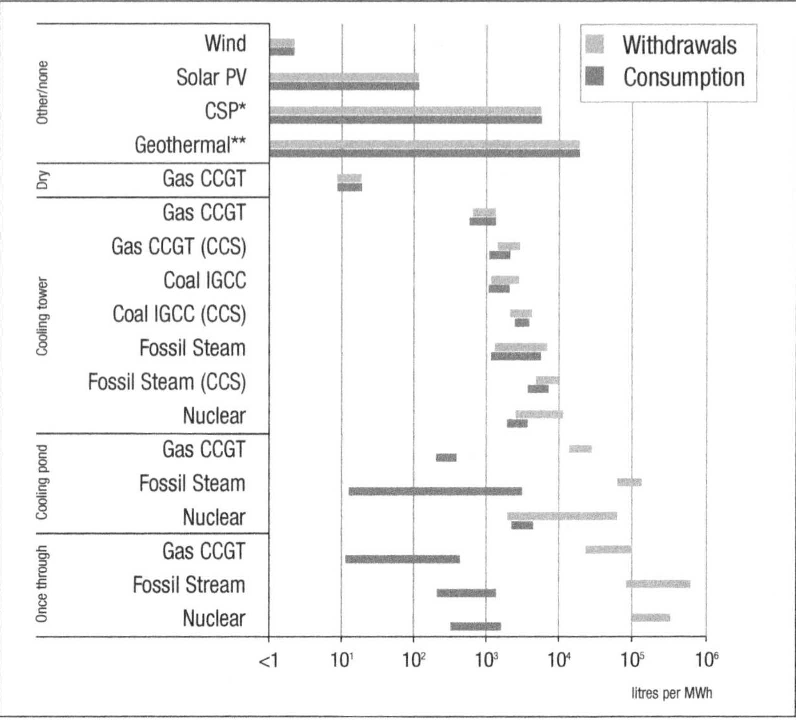

Source: Hoffman, A., Olsson, G., Lindstrom, A. 2014

Table 3: Average water consumption by transportation fuel (gal*/million BTUs)

	Raw materials	Transformation
Oil (traditional)	1.4	12.5
Natural gas (as on land)	0	2
Unconventional natural gas (shale)	12.5	2
Oil sands	260	12.5
Enhanced oil recovery	1,257	172
Biofuels (irrigated corn)	15,750	9
Biofuels (irrigated soy)	44,500	9

*One gallon = 3.78 litres

Source: Glassman, *et al.* 2011

The Nexus From a Water Perspective

From a water perspective, energy consumption and loss of energy through the water value chain and the impact of energy outages, specifically electricity, on the water supply chain, are key (Table 4). Water weighs 8.35 pounds per gallon and therefore requires a significant amount of energy to lift (Water in the West, Undated). The energy intensity of each stage of the water-use cycle can vary significantly (Table 5) depending on the source of the water, topography between the source and places of use, stage of the water supply chain, technology deployed, condition of infrastructure, and the quality of the water being treated.

Table 4: Power outage impacts on the water supply chain

Stage of water supply	Impacts
Abstraction	• Adverse impact on pumps, equipment and telemetry devices • Problem in extracting water • Adverse impact on users of small-scale abstraction schemes such as boreholes
Water treatment	• Adverse impact on equipment, pumps, telemetry devices and dosing apparatus • Difficulty in transporting water • Negative impact on water quality due to non-functional water treatment processes • Negative impact on water treatment facility by way of revenue loss, reduced operational capacity, increased labour costs, water wastage, increased pump start-up costs and possible back-up generator costs
Water distribution/reticulation	• Non-operating pumps and telemetry equipment • Adverse impact on water distributed • Higher costs on account of back-up generators, portable water storage tanks for local communities, portable sewage spill bins and sewage spill clean-up costs
Wastewater treatment	• Non-operating pumps and telemetry devices • Limited control of treatment stops and sewage flows • Higher costs on account of equipment damage, possible back-up generator costs, portable sewage spill bin costs, increased labour costs and increased pump start-up costs

Source: Winter 2011

Table 5: Range of energy intensities for water-use cycle segments

	Range of energy intensity (kWh/MG*)	
Water-use cycle segments	Low	High
Water-supply and conveyance	0	14,000
Water treatment	100	16,000
Water distribution	700	1,200
Waste-water collection and treatment	1,100	4,600
Waste-water discharge	0	400
Recycled water treatment and distribution	400	1,200

*MG – million gallons

Source: California Energy Commission (2005)

A comparison of water source options available to water utilities suggests that, with energy intensity ranging from 1,000 kWh/MG to 500,000 kWh/MG, desalination is often the most energy-intensive water option and water recycling is often the least energy-intensive option for utilities (Figure 5). Some estimates suggest that desalination could be 10 times more energy intensive than accessing local water resources (Hoff, 2011). Energy is also the largest single variable cost for a desalination plant. It is estimated that energy cost varies from one-third to more than one-half the total cost of desalinised water (Chaudhry, 2003 in Water in the West, 2013). Further, a 25 per cent increase in energy cost could potentially raise the cost of produced water by 11 per cent and 15 per cent for reverse osmosis and thermal plants, respectively (Cooley, *et al.* 2006 in Water in the West, 2013).

Figure 5: Energy Intensity of Water Supply Options for Inland Empire Utility Agency, US

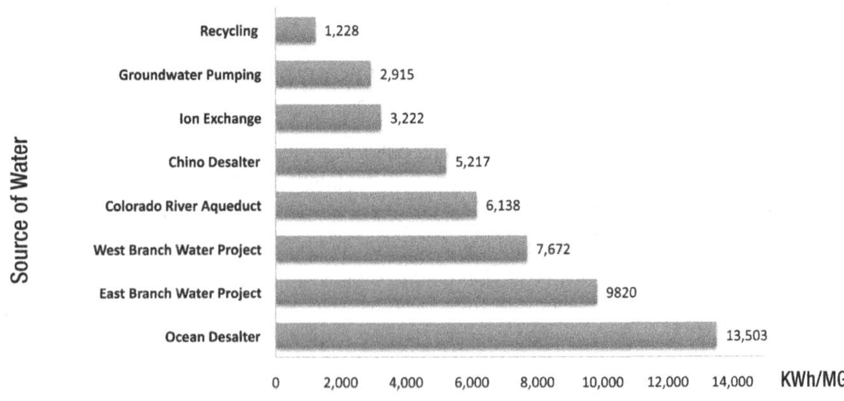

Source: Water in the West, 2013

The energy intensity of a desalination plant depends on the quality (i.e. saltiness) and volume of water being desalted and the technology deployed. Similarly, the energy intensity of recycled water depends on the quality of wastewater being recycled and on the end use of this water. The latter determines the standards to which water must be treated and, therefore, the process and technology to be deployed for collecting, treating and disposing wastewater. Recycled water intended to be potable needs to be treated to high-quality standards, necessitating advanced treatment technologies that often have higher energy requirements (Gulati, 2014a). The distribution of recycled water often has a higher energy cost than the distribution of potable water, since wastewater facilities are typically sited at lower elevations to take advantage of gravity (Water in the West, 2013).

The next important aspect is the loss of energy in the water cycle and its impact on the cost of water supply and services. Energy is lost in the water cycle for reasons such as leaks in water conveyance systems or inefficient use of water; inefficient pump stations due to poor design; old pipes with high head loss; and bottlenecks in the supply network (Feldman, 2009). Leaking water distribution systems and lost water mean that utilities have to produce a greater volume of treated water, implying increased energy consumption for abstraction, treatment and distribution (Gulati, 2014a). It is estimated that, when global water-loss average is approximated at 30 per cent, the same quantum of energy is lost (Feldman, 2009). This in turn suggests that the potential for energy savings in the water cycle can be as high as 20 to 30 per cent of current consumption (Gulati, 2014a).

In case of energy outages, the impact on the water supply chain depends on plant characteristics and availability of back-up power (Winter 2011). Pumping is most vulnerable to electricity outages in the water supply chain. Energy outages also affect water security for end users due to the adverse impacts on abstraction, distribution or water treatment points in the supply chain (Winter 2011).

THE NEXUS PERSPECTIVE OF ENERGY AND WATER RELATED CHALLENGES IN SOUTH AFRICA

The energy sector in South Africa is highly reliant on coal. Nearly 86 per cent of the electricity in the country is produced from coal (Figure 6) on the back of relatively water-intensive, wet-cooled coal power stations. Not surprisingly, there is codependence on electricity production and water use (Figure 7) and a shortage of water can affect electricity availability. The water requirement of electricity generation technologies in the context of South Africa is provided in Table 6. In 2010, wet-cooled coal power stations represented approximately 78 per cent of the country's power generation, while consuming 98 per cent of the water requirements of Eskom (Eskom, 2011).

Figure 6: South Africa's electricity generation capacity (2011)

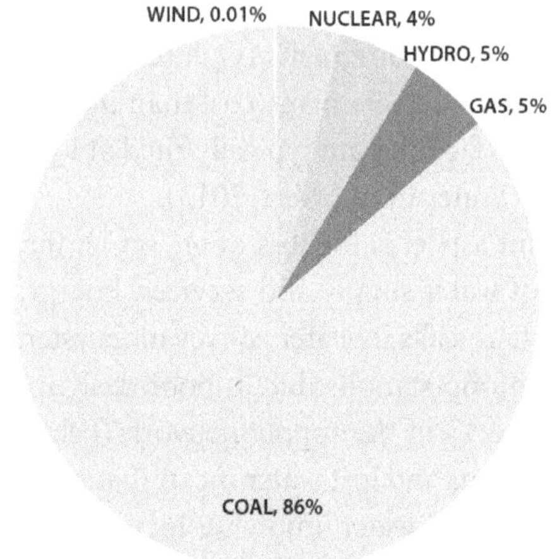

Source: Eskom

Table 6: Water use by electricity-generation technology type for South Africa

Technology type	Water Use (ℓ/kWh)
Wet-cooled coal (existing)	1.15–2.30
Wet-cooled coal (future)#	2.12–2.80
Dry-cooled coal (existing)	0.11
Dry-cooled coal (future)*	0.36
Nuclear	0.055
Open cycle gas turbine	0.01
Combined cycle gas turbine	0.25
Solar photovoltaic (PV)	0.01
Concentrated solar power (dry-cooled)	0.34
Wind	0

Refers to committed and uncommitted future capacity

* Includes flue-gas desulphurisation (FGD) technology

Source: Eskom, 2011

Figure 7: Relationship between water consumption and energy production by Eskom from 1994 to 2005

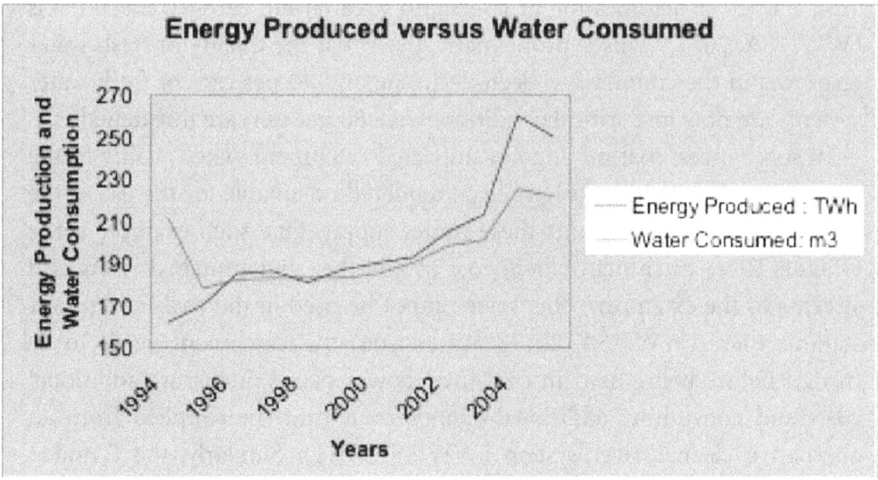

Source: Eskom, 2008 in Winter (2011)

This high dependency of the country's energy sector on water is of paramount concern given the grim situation of water resources. Out of the 19 water management areas (WMAs), five were already experiencing water shortages by 2000, 10 areas broke even and only four enjoyed a surplus of water. It is estimated that by 2025 South Africa's water shortfall would be 1.7 per cent based on current water usage. Shortage of water could therefore affect electricity generation.

The problem can be put in perspective when combined with the fact that most of the country's coal-based power plants are within regions partially or severely constrained in terms of water (Pouris and Thopil, 2015). Existing plants such as Camden, Komati and Grootvlei are located in the Olifants and Inkomati WMAs that are severely constrained while new coal-fired power plants, Medupi and Kusile, are located in the moderately constrained WMA of Limpopo and the severely constrained Olifants WMA, respectively (ibid.).

Various measures have been implemented in the country to conserve water at power stations. This includes a shift to dry-cooling technology for coal-fired power stations, which have 5 to 10 per cent of the water requirements of wet-cooled stations (Eskom, 2011). Nevertheless, these power stations are still 100 per cent dependent on water (Eskom 2011) and a rising demand for energy in the future has the potential to significantly increase water consumption.

The dependence on coal also raises concerns in terms of the impact of its extraction on water quality. Acid mine drainage (AMD) from coal-mining areas has led to acidification of rivers and streams and elevated metal levels (WWF-SA, 2011). This is problematic given that the quality of fresh water resources in the country has been declining and 40 per cent of fresh water systems are now in a critical condition while 80 per cent are threatened.

In some cases, coal mining has impacted catchment water quality to the extent that the catchments have been rendered unsuitable for the use of the very coal-fired power plants these mines supply. One such example is the Olifants River catchment where coal mining has contaminated rivers and streams to the extent that the water cannot be used in the coal-fired power stations there (WWF-SA, 2011). Water from this catchment needs to be treated before being used in coal-fired power plants, incurring additional costs and consuming additional energy, or it must be supplied from an alternative cleaner river system (WWF-SA, 011). Similarly, the Camden power station in Mpumalanga is reliant on inter-basin transfers from the unimpacted Usutu River system originating in Enkangala for water that is fit to use in the power station (WWF-SA, 2011).

Although legislative changes have ensured that mine water is integrally considered in the mining process, and have led to good practices such as the eMalahleni water purification plant situated in the Witbank coalfields of the Mpumalanga province, turning mine effluent into a usable resource (WWF-SA, 2011) across the industry requires significant technical and financial resources that are often beyond the reach of smaller mining companies.

Heavy reliance on coal also makes the electricity sector account for half the country's greenhouse gas emissions. The electricity system is thus a key point of intervention from the environmental perspective. Measures to this end include an increasing share of renewable energy (RE) in the electricity generation capacity, and the development of carbon capture and storage (CCS) technology that has the potential to achieve emission reductions of between 80 and 85 per cent (Eskom, 2011), and installation of flue-gas desulfurisation (FGD) technology at coal-fired power stations for meeting local air-quality standards. These measures could have a significant bearing on the energy sector's water requirements.

Some RE technologies, such as concentrated solar power (CSP), are water-intensive technologies, even though they place a lower demand on water resources compared to coal-fired power generation. As with coal power plants, CSP plants are also located, or are proposed to be located, in areas facing water stress. This means that local water availability for these plants could potentially be a problem (Gulati, 2014a). This water requirement can be mitigated through the same dry-cooling technology as coal-fired plants (Gulati, 2014a). The bidding process deployed to select the CSP projects to be developed in the country does not provide for a preferential payment to CSP with dry cooling. Since the cost of implementing this technology is about five times greater than the regular wet-cooling technology (Gulati, 2014a), it is important that CSP with dry cooling be incentivised in the process.

Carbon capture and storage (CCS) technology could also increase the water consumption of power plants between 46 and 90 per cent depending on the technology deployed (Eskom, 2011). Similarly, retrofitting existing coal-fired power stations with FGD and installing FGD at all new coal-fired power stations could dramatically increase water requirements for the electricity sector (Figure 8), although the water requirements in 2030 are expected to be lower than 2011 requirements (Eskom, 2011).

Figure 8: Water requirement implications of FGD retrofitting and installations

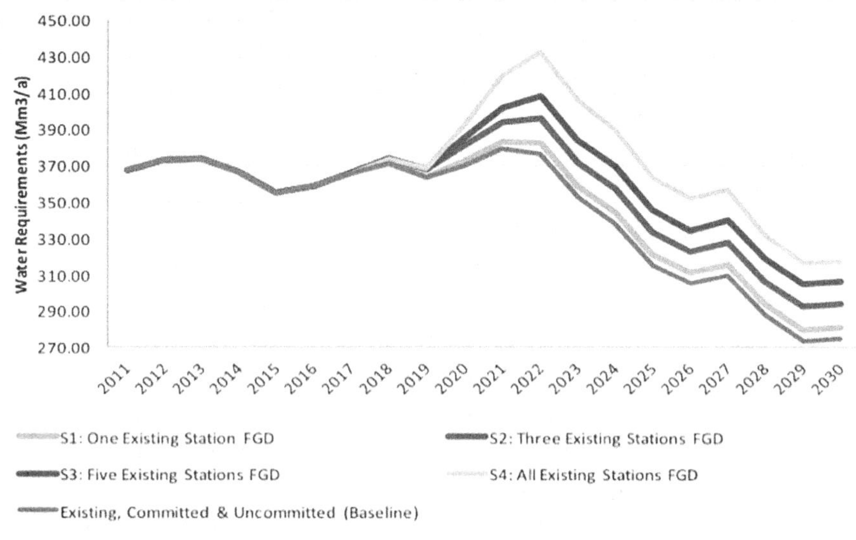

Source: Eskom (2011)

Eskom's modelling of future water requirements for the most likely power-generation scenarios in the country show that 1) water requirements in 2030 would increase by 23 Mm^3/annum with coal replacing the planned nuclear capacity and FGD being installed on only new coal capacity; 2) 42.5 Mm^3/annum with existing plants being decommissioned as planned and FGD being installed at all existing, committed and uncommitted power stations; and 3) 173.7 Mm^3/annum with a generation-capacity gap, no existing plants being decommissioned and FGD being installed at all existing and future power stations. The third scenario has the biggest potential impact on water requirements in the future and could therefore pose serious trade-offs for water allocation to agriculture if it materialises (Gulati, 2014a).

The issue of nuclear is of significance here. The Integrated Resource Plan (IRP) 2030 plans for the procurement of 9.6 gigawatts (GW) of nuclear power. On the water-use side, nuclear power plants use large quantities of water for steam production and for cooling. Some nuclear power plants remove large quantities of water from a lake or river. On the water-quality side, the build-up of heavy metals and salts in the water used, as well as the higher temperature of the water discharged from nuclear power plants, can negatively affect water quality. Waste generated from uranium mining operations can contaminate groundwater and surface water resources with heavy metals and even traces of radioactive uranium.

Two other lower carbon technologies also merit discussion: gas-to-liquids (GTL) and hydraulic fracturing for shale gas. GTL is the conversion of natural gas into petrol distillates such as transportation fuel or other chemicals. Once again, a wide range of values for consumptive water use can be associated with this technology. The water-intensity average of GTL is 42 gallons per MMBTU and ranges from 19 to 86 gal/MMBTU.

Hydraulic fracturing is a technology used to harness shale gas and involves pumping fluid composed of water, proppants and chemicals into the ground at very high pressure. The pressurised water and chemicals create and enlarge cracks in the shale formation, increasing its permeability by 100- to 1,000-fold and allowing hydrocarbons to flow more easily to the wellbore (Reig, Luo and Proctor, 2014). Shale gas resources in the country are pegged at around 485 trillion cubic feet (Tcf), though the economically recoverable reserve is yet to be determined (Fakir, 2015). The volume of water required for hydraulic fracturing depends on the nature of the fractures that have to be performed. Based on experience in the United States, drilling a single well can require between 0.2 million and 2.5 million litres of water and hydraulic fracturing of a well can require between 7 million and 23 million litres of water, 25 per cent to 90 per cent of which might be consumptive use (Reig, Luo and Proctor, 2014). The water required by a single well can be roughly equal to the water consumed by New York City in seven minutes, or by a 1,000-megawatt coal-fired power plant in 12 hours (Reig, Luo and Proctor, 2014).

Limited experience with hydraulic fracturing and the wide range of values for consumptive water use associated with it indicate the high levels of uncertainty about its possible impacts on freshwater availability, although nationally the effects of water consumption for fracking may not be significant. That said, given that South Africa is a water-stressed country, the vast water quantities needed over the lifespan of a shale gas well, where water is used to fracture rock under high pressure, may pile further stress on local fresh water sources (SIWI, 2014).

The problem with the use of potable water is that a quantity of it is permanently lost and what is returned as flow-back and produced water (essentially wastewater from fracturing) is a mix of the originally pumped fresh water (which is now contaminated) and formation water (water rich in brine from the targeted shale gas-rich rock) (Fakir, 2015). Water that is not recycled or reused will incur an opportunity cost given the local water stress (Fakir, 2015). If properly treated, returned water can be reused in other fracking operations. But treatment methods for the returned water are often inadequate to achieve a potable water standard. If improved treatment procedures are developed, it will most likely be at considerable cost. Water quality is also under threat from fracking (SIWI, 2014). The risk of ground water contamination can be high if stringent standards are not adhered to for drilling of the borehole, installing the well casings and cementing them in place (Fakir, 2015).

The green economy demands not only a low-carbon economy but also a low-water-use economy. Therefore, the low-carbon transition needs to be planned with due consideration to the implications for water requirements. Otherwise, the low-carbon economy could risk constraining the country's energy supply further and go against the tenets of the green economy.

The country is currently facing an electricity crisis. This crisis is expected to continue at least over the next two to three years. This is likely to aggravate the country's water crisis, which is already projected to be under stress with the growing population. There already exist examples of how the energy shortage in the country has led to plummeting reliability of water systems while threatening public health and safety. During the recurring rolling blackouts of 2007 and 2008, municipalities were unable to provide water and wastewater services. While Cederberg and the City of Cape Town experienced higher costs on account of damage to equipment and the cost of back-up services, Howick in KwaZulu-Natal and Zandvlei in the Western Cape experienced adverse health and environmental impacts, respectively (Winter, 2011).

Impending water scarcity and deteriorating water quality would speed up the exploration of new processes or technologies to access or treat existing water resources to make them usable. Desalination and wastewater treatment could be energy intensive (Table 7), and the shortage of energy or rising energy prices could either prohibit the feasibility of these options, or involve trade-offs between water and energy security.

Table 7: Energy consumption range for the South African water supply chain (in kWh/M)

Process	Minimum	Maximum
Abstraction	0	100
Distribution	0	350
Water treatment	150	650
Reticulation	0	350
Waste water treatment	200	1,800

Source: Winter (2011)

In the past, the water sector, as has been the case with every sector in the country, has not given precedence to electricity efficiency measures due to historically low electricity tariffs. This situation started changing since 2008 on account of electricity shortages and rising electricity prices. Average electricity prices increased by 24 per cent per annum between 2008 and 2012. While further annual average increases of seven per cent were estimated until 2019, actual increases in many years have been well over seven per cent. For example, in 2015 average electricity prices rose by 12 per cent. There is no doubt that these price hikes are having a substantial effect on the actual cost of service of water utilities. Thus, there is a need to ensure that the selection of technology must consider electricity costs, and technologies should be as cost effective as possible and should pursue energy-efficiency optimisation in waterworks (Scheepers and Van der Merwe-Botha, 2013).

On the water and wastewater treatment side, the inability of existing water resources to dilute pollutants would necessitate that raw water will increasingly have to be treated, or standards for water treatment will have to be progressively increased (Gulati, 2014a). Wastewater management and treatment has, in fact, long been considered an important instrument for public health and the control of pathogens, and municipalities are consistently faced with stricter effluent discharge standards (Scheepers and Van der Merwe-Botha, 2013). Currently, only 76 per cent of Water Service Authorities (WSAs) treat the raw water supplied to end users and the level of treatment varies considerably amongst them (Winter, 2011). With all WSAs treating water to 100 per cent, energy consumption and related costs could increase substantially, necessitating higher municipal service charges.

This does not imply that there is no understanding of the link between energy and water amongst policymakers and planners (Gulati, 2014a). Water is recognised as a key constraint in the planning of future electricity generation capacity under IRP 2010–2030. The Biofuels Industrial Strategy, which provides the framework and incentives for the development of biofuels, initially proposed a five per cent penetration level of biofuels in the national liquid fuels pool. This was reduced to two per cent after the National Treasury expressed concerns over the water requirement implications of such a mandate.

However, policies have deficiencies. For example, IRP 2030 only considers water usage under the decision-making criteria. It does not consider either the risks of potential water scarcity for the planned generation capacity and electricity supply or the electricity sector's ability to provide reliable and sustainable energy supply in the event of water insecurity (Gulati, 2014a). Similarly, the mandate for all new coal power plants to utilise dry-cooling technology can be short-sighted if it omits an effective and long-term solution for a low-water-intensive electricity system (ibid.).

POLICY RECOMMENDATIONS AND CONCLUSION

Understanding the EWN and managing energy and water resources in a comprehensive and systematic way is critically important to meeting the basic needs of these resources and achieving the green economy objectives. It provides the opportunity for integrated planning that is key to assessing pathways that are affordable in the short-term and sustainable for the long-term, thereby meeting the concept of the green economy. In fact, there are risks that energy and water policies developed in isolation to increase efficiency in one sector may not only be creating additional demand in the other sector but also creating risks pertaining to broader allied economic activity and lowering the resilience of relevant social systems. Below are some recommendations that would facilitate the cooperative and adaptive management of these vital resources and help to achieve the green economy goals for South Africa.

Addressing Data Gaps

There is a clear need to improve the quality of the data on energy-for-water and water-for-energy use. The real understanding of the energy-water nexus and its impact on green economy objectives, such as resource security and economic and social development, will be possible only with reliable, current and comprehensive national and local-level data. Improved availability and accuracy of data will facilitate informed decision-making; prioritise investments in both energy and water infrastructure; and lead to better water and energy use practices. Examples of data requirements include water usage for liquid fuel production; water usage for exploration of new energy sources; the impact of new energy sources on water quality; energy use by new water sources; water usage patterns for new coal technologies; mapping of water usage of coal power plants in relation to water inventories in water deficient WMAs; the impact of low-carbon technologies on water usage; and water usage patterns in renewable energy technologies (Pouris and Thopil, 2015 and Gulati, 2014a). Significant attention needs to be paid to improvements in industry reporting, data collection and the sharing of data.

Integrated Resource Planning and Management

The EWN needs to be better recognised in policies, planning and related regulations, and laws dealing with the management and development of water resources and energy systems. This will optimise potential benefits, provide the right business and investment environment, protect the environment, and further the basic philosophy of the green economy.

This process could start with a review of existing policies to identify disincentives or positive synergies to integrated planning and management and sustainable supply of the two resources. Sectoral policies and plans for both water and energy need to better account for the multiple interrelationships of the two resources, as well as the impact of these interrelationships on different sectors of the economy.

To begin with, water security needs to be central to the energy-planning debate. In the event of regional integration of energy supplies and possible regional hydropower procurement, the energy-water nexus could become an important debate. Then policies should support the upscaling of new energy-efficient water technologies and water-efficient energy technologies, even though these may be expensive and deter energy technologies that pose risks for water availability or quality. In the case of new and emerging technologies, these must not be allowed unless the true scale of water impacts can be estimated.

There will be practical difficulties in achieving such integrated planning as water-resource planning is carried out at the level of river basins and energy planning is done at a national level. Nonetheless, coordinated planning will help meet the objectives of the green economy.

Coordinated energy efficiency and water conservation programmes could help garner savings of both water and energy, and synergistic energy and water production could enhance both water and energy security. For example, power plants and desalination or water treatment facilities could be co-sited and power plants could provide electricity at a preferential rate to these water-related facilities.

For example, minimising water loss through an active leakage reduction programme will reduce the energy wastage embedded in the lost water. Waste heat from power plants can be used in some desalination cycles, and biogas from wastewater treatment plants can be used for combined heat and power production. This is a relatively untapped energy source in the country.

Efficiency Improvement

There is an urgent need to prioritise reduction in leakage and energy efficiency in the water sector. Measures that can be considered to this end include water pressure management (as water leakage is driven by water pressure), sludge management activities such as aeration efficiency improvement, varying operating conditions, using chemical pre-treatment (this could lead to a potential electricity saving of 250–280 MWh/year), and optimisation of the operation of the distribution system by pumping at off-peak periods while ensuring minimum emergency levels in all reservoirs (Swartz, van der Merwe-Botha and Freese, 2013). The latter will not only effect energy savings in terms of electrical energy used for pumping, but may also delay upgrades to the water distribution system (ibid.). Energy audits form the key to the identification of energy savings and therefore need to be mandated for water and wastewater systems.

Appropriate Pricing of Water and Energy Resources

Appropriate resource pricing can play an important role in driving conservation and innovation in management and use of both resources. Underpricing of resources encourages wasteful consumption of resources and misses a valuable opportunity to secure resources for effective protection and management of water. Energy and water subsidies help drive the cycle of inefficient and energy-intensive water use by hiding the true resource costs (Wolff, *et al.* 2004 in Water in the West, 2013). Of course, principles of equity would need to be considered and applied when higher pricing of water and energy resources is implemented. There is also a need to offer better incentives or to reconsider subsidies to manage the risks posed by the one resource to the other. For example, CSP project developers need to be given incentives to deploy the water-saving, dry-cooling technology.

Increased Funding for Research and Development (R&D)

The EWN presents the opportunity to develop more efficient technologies, practice cost-effective approaches to using lower-quality, non-traditional sources of water to supplement or replace fresh water for cooling and other power plant needs. Future R&D should focus on the following areas:

- energy sources that can meet future water needs sustainably, specifically for the array of water-scarce areas in the country;
- technical solutions that successfully couple energy and water generation;
- reducing water use in thermal power generation through advanced cooling technologies, scrubbing, innovative source-water intake designs, use of non-traditional waters and increased power-plant efficiencies;
- treatment of returned water from fracking operations to enable its reuse;
- applications and treatment methods for non-traditional or lower quality water sources such as saline or brackish water specifically aimed at providing alternative or supplementary sources of water for energy generation and for uses that could supplement water resources in water-stressed areas; and
- determining the manner in which the state and the costs of the existing and future power supply affect the costs of water-related technologies and the capacity to deliver water services in the country.

Creating Consumer Awareness

While the electricity tariff hikes and rising oil prices in recent years have created an awareness of energy consumption amongst consumers, awareness remains rather low on the water-use front. Urgent steps need to be taken by the government, and the water companies need to take steps to increase consumer awareness of water use. Tools such as the certification and labelling of consumer products to reflect embedded water and energy use in their manufacture or usage need to be mandated.

References

Eskom. 2011. 'Eskom's submission to the DWA for the National Water Resources Strategy review'. Johannesburg.

Fakir, S. 2015. 'Framework to assess the economic reality of shale gas in South Africa'. WWF-SA, South Africa.

Greenpeace Africa. 2012. *Water hungry coal. Burning South Africa's water to produce electricity*. Johannesburg.

Gulati, M. 2014a. 'Understanding the Food Energy Water Nexus: Through the Energy and Water Lens'. WWF-SA, South Africa.

Gulati, M. 2014b. 'Financing the Green Economy Transition in Africa. Greening the Continent: Reflections on Low-Carbon Development Pathways', *Perspectives Africa*, Al-Zubaidi, L., Luckscheiter, L. and Omari, K. (eds.), Issue 2, September. Cape Town: Heinrich-Boll-Stiftung, Regional Office Southern Africa.

Hoffman, A., Olsson, G. and Lindstrom, A. 2014. *Shale Gas and Hydraulic Fracturing: Framing the Water Issue*. Report No. 34. Stockholm: SIWI.

Paul, R. 2003. 'Sectoral trends in the water sector (technology, policy and poverty) in South Asia', *South Asia Conference on Technologies for Poverty Reduction*, 10–11 October. New Delhi.

Pouris, A. and Thopil, G. A. 2015. *Long-term Forecasts of Water Usage for Electricity Generation: South Africa 2030*. Report No. 2383/1/14. ISBN 978-1-4312-0646-9. Pretoria: Water Research Commission.

PricewaterhouseCoopers Inc. (PwC). 2011. 'Water, food, energy and the green economy'.

Reif, P., Luo, T. and Proctor, J. N. 2014. *Global Shale Gas Development: Water Availability and Business Risks*. Washington, DC: World Resources Institute.

Swartz, C. D., van der Merwe-Botha, M. and Freese, S. D. 2013. *Energy Efficiency in the South African Water Industry: A Compendium of Best Practices and Case Studies*. Report No. TT 565/13. ISBN No: 978-1-4312-0430-4. Pretoria: Water Research Commission.

United Nations Economic Commission for Africa (UNECA). 2013. 'Enhancing Energy Access and Security in Eastern Africa'. Draft background report. 17th Meeting of the

Intergovernmental Committee of Experts, 18–22 February 2013. Kampala, Uganda.

United Nations Environmental Programme (UNEP). 2011. *Towards a Green Economy: Pathways to Sustainable Development and Poverty Eradication*. www.unep.org/greeneconomy. ISBN: 978-92-807-3143-9.

US Department of Energy (US DoE). 2006. *Energy Demands on Water Resources*, report to Congress on the Interdependency of Energy & Water. Washington DC.

'Water in the West'. Undated. *Water and Energy Nexus: A Literature Review*. Stanford, California: Stanford University.

Winter D. 2011. *Power Outages and their Impact on South Africa's Water and Wastewater Sectors*. Report to the Water Research Commission. WRC Report No. KV 267/11 ISBN 978-1-4312-0101-3.

CHAPTER 8

Waste Re-use: Oil Extraction from Waste Tyres and Improvement of the Waste Tyre Industry

T. J. Pilusa and Edison Muzenda

EXECUTIVE SUMMARY

The work that went into this chapter was aimed at researching and identifying technologically feasible and economically justifiable processes and pathways for the utilisation of waste tyres in South Africa. Drawing from the objectives of the Recycling and Economic Development Initiative of South Africa (REDISA), this chapter seeks to motivate the utilisation and unlocking of the potential of waste tyres for energy and material recovery in South Africa. The chapter focuses on the economic feasibility of waste tyre treatment through pyrolysis in Gauteng. The authors also evaluate the role of existing policy instruments, transition towards a low-carbon economy as well as optimal energy strategies influencing and promoting efforts towards integrated waste tyre management in South Africa.

In line with the National Green Economy Strategy, West Rand Green IQ and the Gauteng Economic Strategy, in particular the Gauteng Integrated Energy Strategy, Gauteng needs to explore alternatives for both energy sources and waste treatment technologies. The successful implementation of an efficient Waste Tyre-to-Energy and Material Recovery concept will contribute to overall sustainability by improving waste collection management, the sustainable utilisation of waste tyres, improving the environment by increasing resource efficiency and energy poverty reduction, as well as promoting economic development such as the optimisation of green jobs.

The South African government recognises the conversion of waste to energy in its plans and future strategies (Department of Environmental Affairs, 2010).

The end-life of automotive tyres is associated with various environmental challenges. Combustion of waste tyres emits significant amounts of toxic gases and particulate matter, which contribute significantly to air pollution. Illegal re-treading of tyres increases the risks of road accidents and the stockpiling of waste tyres may results in fire hazards and air pollution. Waste tyre stockpiles create breeding grounds for vermin carrying insects and mosquitoes, which threaten human and animal health. Landfill disposal is not viable due to the non-biodegradability of the tyres and the large volumes they occupy in landfill sites. They are also difficult to compact after disposal. Waste tyre treatment through pyrolysis results in four revenue streams: tyre-derived fuel, refined carbon black, high tensile steel and sodium sulphite.

This study focused on:

- the economics of waste tyre pyrolysis technology in the South African context;
- quality of the products derived from this process;
- methods of beneficiating these products for local and export markets;
- potential uses of value-add products derived therefrom; and
- socio-economic benefits of the combined approach.

An economic model was developed to assess the viability of pyrolysis technology as an alternative treatment method for waste tyres using Gauteng Province in South Africa as a case study. This chapter focuses on the factors that influence the viability of using pyrolysis technology as a treatment process for waste tyres with the aim of producing alternative fuel and other high-value products. A financial model was formulated to evaluate the economic feasibility of such technology as an alternative disposal method. Pyrolysis technology becomes more viable when there are guaranteed product off-takes at a given price. Further processing of the crude tyre oil and carbon black is important for production of consistent quality products.

Gauteng Province alone generates approximately 48,077tpa of waste tyres of which four per cent is already allocated to three waste tyre pyrolysis recyclers in the province (REDISA, 2014). These facilities are equipped with 10tpd modular batch reactors. One of the facilities is in the process of installing three additional reactors to increase the treatment capacity to 40tpd, whereas the other facility has two 10tpd reactors but is limited to process 10tpd since it is operating without an environmental impact assessment. If it is assumed that the balance will also be treated via pyrolysis technology, a capital injection of R129 million will be required to set up three additional 30tpd waste tyres pyrolysis treatment facilities to treat the balance of waste tyres. Based on the financial model evaluated, a complete 30tpd treatment facility, complete with product refinery modules, has a potential return on investment and gross margin of 27.1 per cent and 40.6 per cent, respectively. Each facility can produce up to four million litres per annum of refined tyre-derived fuel at a cost of R5.46 per litre and other secondary value-add products for local and export markets.

The chapter presents an economic evaluation of waste tyre pyrolysis technology, taking into account primary products, 'upgrade/beneficiation/refinery', efficient plant energy re-utilisation and minimal environmental impact. Production of alternative fuels from this technology is profitable and feasible depending on the product's demand at the minimum selling price. Successful application of the liquid fuels produced from this process will create an attractive market for fuel at a higher selling price but lower than the conventional petroleum liquid fuels such as diesel. Based on the findings of this study, the Recycling and Economic Development Initiative of South Africa's (REDISA) Integrated Waste Tyre Management Plan (IWTMP), established in terms of the Regulation 6(3) of the Waste Tyre Regulations in accordance to the National Environmental Management: Waste Act, 2008 (Act No. 59 of 2008), must be supported as an integrated waste tyre management option in South Africa. The REDISA plan and state-owned entities, such as the South African National Energy Development Institute (SANEDI), are making a significant contribution towards meeting government objectives as defined in the National Development Plan (NDP) 2030, such as economy and development, environmental sustainability and resilience, and transforming human settlements. Therefore, important strategic efforts such as those by SANEDI, REDISA and the Mapungubwe Institute for Strategic Reflection (MISTRA) can assist the South African government to optimise research strategies and policies that promote transition to the low-carbon economy through investment in scientific research and technological advancement as well as human capacity development.

BACKGROUND

Waste Tyre Quantities

It is estimated that 1.5 billion tyres are produced globally per year and the majority of these eventually end up as waste tyres contributing a significant portion of the solid waste stream. The European Union generated an estimated 4.5 million tonnes of new tyres in 2010, with 289 million tyres being replaced per year (ETRMA, 2011). In the United States, approximately 500 million waste tyres were generated in 2007, with about 128 million used tyres already stockpiled (Berrueco, *et al.* 2005). In Australia, around 52.5 million tyres reached their end of life between 2007 and 2008. Approximately 64 per cent of these tyres were landfilled, or illegally dumped or stockpiled, while only 13 per cent were recycled (ETRMA, 2011).

Disposal of waste tyres and rubber products in South Africa is becoming a growing concern since these products do not decompose easily and take up a significant portion of disposal site space. There are an estimated 60 million scrap tyres disposed across South Africa, with an additional 11 million adding to the stock pile in 2008 (Mahlangu, 2009). In 2011 alone, 7.25 million new tyres, excluding mining tyres and belting, were sold in South Africa: approximately 55 per cent of these tyres were sold in Gauteng Province. The majority of these tyres are expected to add to the existing waste tyre matrix (REDISA, 2013).

It is well known that tyres possess highly volatile and low ash contents with a heating value greater than that of coal and biomass. These properties make them ideal materials for thermal processes like pyrolysis, gasification and liquefaction (Bridgwater, 2003; Kiran, *et al.* 2000; Morris and Waldheim, 1998). In 2009, the South African Department of Environmental Affairs reacted to the potential danger to the environment caused by waste tyres by demanding that the tyre industry be the first in South Africa to develop an industry waste management plan. The Recycling and Economic Development Initiative of South Africa (REDISA) (Furlonger, 2014) drives the plan, which became effective in 2013. The initiative had collected about 28,800 tonnes of old tyres by June 2014 (Furlonger, 2014). This shows a challenge in the management of waste tyres as about 240,000 tonnes of new tyres enter the market every year in South Africa. The plan relies on independent transporters delivering old tyres to regional storage depots from where they are passed on to recyclers (Furlonger, 2014). Although REDISA is making a significant contribution, a huge market for waste tyres is required to address this growing environmental challenge. Under the new plan, tyre manufacturers and importers are paying a levy of R2,300/tonne, plus 14 per cent value added tax for any new tyre they introduce to South African roads (REDISA, 2014). The funds raised will enable a move to kick-start a new industry around the collection, transport and disposal of every waste tyre (Molewa, 2012). By creating income generating opportunities, and the researching of new recycling methods, the new legislation will dramatically reduce the quantity of waste tyres in South Africa. The Recycling and Economic Development Initiative of South Africa, as gazetted by the Water and Environmental Affairs Minister on 23 July 2012, has been well received. All tyre producers registered with REDISA must immediately comply with the approved REDISA Integrated Waste Tyre Management Plan (IWTMP) in terms of Regulation 6(3) of the Waste Tyre Regulations in accordance with the National Environmental Management: Waste Act, 2008 (Act No. 59 of 2008). Pyrolysis, gasification, and liquefaction are technologies that could be used to divert a significant portion of the scrap tyres currently being landfilled.

Waste Tyre Pyrolysis Technology

There has been great interest in alternative treatment processes for waste tyres, amongst which is the use of pyrolysis technology (Sienkiewicz, *et al.* 2012). Pyrolysis is the thermal degradation of the organic components of the tyres at typical process temperatures of 400°C to produce oil, gas and char product, in addition to the recovery of the steel. The yields and quality of these products are influenced by the reactor type used to pyrolyse the feedstock (Bhatt and Patel, 2012). The fuel oil produced from this process is a complex chemical mixture containing aliphatic, aromatic, heteroatom and polar fractions (Pilusa, *et al.* 2013). The fuel characteristics show that it has very similar properties to diesel fuel and can be used in test furnaces and internal combustion engines (Murugan, *et al.* 1998).

The oil may be used directly as a fuel, added to petroleum refinery stocks, upgraded using catalysts to a premium grade fuel, or used as a chemical feedstock (Islam, *et al.* 2010). The gases from tyre pyrolysis are typically composed of C1–C4 hydrocarbons and hydrogen with a high calorific value of sufficient energy content to act as fuel to provide the heat for the pyrolysis process (Martinez, *et al.* 2013). The solid char consists of the carbon black filler and also char (Qu, *et al.* 2006). The liquid and gaseous fractions obtained are a valuable fuel source, while the solid fraction (char) has the recovery potential of low-grade carbon black or as carbon adsorbent after applying an activation step (Kalitko, 2008). The uncondensed gaseous fraction has properties close to those of synthesis gas as described by Bajus and Olahova (2011).

The main option for treating waste tyres is as fuel in cement kilns. Other energy recovery options for tyres include use in power plants and co-incineration (Quek, *et al.* 2013). Large portions of tyres are used in material recovery options such as crumb products for the production of rubberised flooring in sports fields and playgrounds, paving block and other rubber applications. A significant proportion of the waste tyres are typically used in civil engineering applications such as road and rail foundations and embankments (Islam, *et al.* 2011). However, there is very little use of waste tyres in civil engineering applications in South Africa.

Liquid Fuel

The total consumption of Heavy Fuel Oil (HFO) in South Africa was 470 million litres in 2007. No latest data is being published due to competition-related concerns. HFO is the lowest cost liquid fuel and prices are usually determined through a negotiation. There are no price regulations for HFO. Globally, the value of HFO has reduced from around the same cost as Brent crude oil to around 70 per cent of the Brent oil price as of 2014. One major reason for the decline in demand is the relatively high sulphur content compared to alternatives. On this basis, the relative price for HFO is at R5.46/litre excluding transportation costs. This is 60 per cent cheaper than commercial diesel fuel and can be sold at a much lower price since there are currently no regulations and levies applied on this specific fuel.

Research has shown that crude tyre-derived oil cannot be used as an alternative fuel in diesel engines without further purifications (Moghaddam, *et al.* 2011). Murugan, *et al.* 2008 investigated combustion of pure crude tyre-derived fuel and its blend with diesel in diesel engines. It was observed that pure tyre-derived fuel cannot be used in diesel engines as an alternative fuel; however, it is possible when it is blended with diesel up to 40 per cent volume. The engine operates efficiently, although there is a high level of sulphur oxide emissions reported due to the fuel quality. An average of 0.23 per cent by volume of sulphur oxide and other emissions was recorded in the engine exhaust emission fuelled with diesel-tyre-derived fuel blend (Murugan, *et al.* 2009). This is significantly high and will pose a serious environmental challenge if the fuel blend is used in diesel-operated vehicles. For this reason, it will be beneficial to use this fuel blend in stationary engines such as diesel generator sets where the resulting emissions can be contained and chemically treated into value-added products.

Emissions

The flue gas emissions resulting from the combustion of HFO-diesel blends and gas-oil in the generator set and reactor heating source are contained and scrubbed through a gas absorption tower using sodium hydroxide solution (Steffens, 2003). Table 1 shows typical flue gas composition from the combustion of heavy fuel oil (HFO). To prevent further air pollution, carbon dioxide can be reacted with sodium hydroxide to form water and sodium carbonate as demonstrated by equation 1. Due to reaction kinetics favouring sodium carbonate formation in the presence of sulphur dioxide, sodium sulphite is formed as a final stable product as shown in equation 2.

Table 1: Tyre-derived Fuel combustion emissions (Tawil, 2008; Schrenk and Berger, 1941)

Component	Concentration (vol. %)	Emission level (g/kWh)
Sulphur Dioxide (SO_2)	0.23%	8.0
Carbon Dioxide (CO_2)	12.5%	296
Nitrogen Oxides (NO_x)	0.003%	0.08
Hydrocarbons (HC)	0.02%	0.18
Carbon Monoxide (CO)	0.01%	0.15
Moisture (H_2O)	10.1%	98
Nitrogen (N_2)	73.6%	1,110
Oxygen (O_2)	3.5%	60

$2\,NaOH + CO_2 \quad Na_2CO_3 + H_2O$ (1)

The initial combination generates sodium-bisulphite ($NaHSO_3$), which is converted to the sulphite by reaction with sodium hydroxide or sodium carbonate (Suresh, *et al.* 2006). The overall reaction is:

$$SO_2 + H_2O + Na_2CO_3 \quad Na_2SO_3 + CO_2 \text{ (2)}$$

Sodium sulphite is primarily used in the pulp and paper industry. It is used in the production of sodium thiosulphate and other applications including froth flotation of ores, oil recovery, food preservatives and dye manufacturing (Weil, *et al.* 1999). The current average market price for sodium sulphite is around R209/tonne.

Carbon Black

The total market for refined carbon black is estimated at around 66,000 tons in 2010. Carbon black is not a single product and large ranges of grades are used in specific applications in order to impart the required properties to the final rubber product. Carbon black grades are determined according to an international ASTM classification. The major multinational tyre companies account for in excess of 80 per cent of the carbon black consumption in South Africa. Non-rubber applications, e.g. pigments in plastics, inks and paints, account for only around five per cent of the market. The average market price for high-grade carbon black is in the order of R6,808/tonne and low grade at R2,812/tonne based on the historical import market (Heckett, *et al.* 2004). The dominant carbon black products in the local market are supplied by Evonik Degussa, which has a local manufacturing plant in Port Elizabeth.

Table 2: Effect of particle size on tyre reinforcement (Norman, 2008)

Particle size nm	Strength
1000–5000	Small reinforcement
<1000	Medium reinforcement
<100	Strongest reinforcement

Particle sizes of commercially used carbon black are very small compared to pyrolitic carbon black. Particle sizes of carbon black are especially important because they are used to categorise carbon black by grades. The average particle size of commercially available rubber-type carbon black grades ranges from about 10 to 500 nanometres (FG, 2010). Small particle sizes of carbon black are favoured in the reinforcement industry because the reinforcement strength increases with a decrease in particle sizes.

Table 3: Classification of activated carbon black by size (Jha, 2008)

Particle Size	Type of Activated Carbon
1–150 μm	Powdered activated carbon
0.5–4 mm	Granular activated carbon
0.8–4 mm	Extruded activated carbon

Carbon black may be used as reinforcing filler in rubber for automotive tyre manufacturing as shown in Table 2, or as activated carbon because of its high carbon content (Jha, 2008). Activated carbon was used mainly for the purification of products of different industries as well as for the purification of drinking water. Jha (2008) classified various size fractions of activated carbon black as shown in Table 3.

A carbon black with a high degree of aggregation is said to have a high 'structure'. Structure is determined by the size and shape of the aggregated primary particles, the number of primary particles per aggregate and their average mass (Navarrete-guijosa and Mostafa, 2007).

The particle size of a carbon black particle is very important. These particles may be categorised into three size ranges. The primary particles have the smallest size, which are typically 13–100nm in diameter depending on the parameters set in carbon black processing (Nerulhuda, 2008). Such primary particles fuse and form in the furnace reactor a second larger level of structure or aggregates having size in the range 200–1000nm. The morphologies of these aggregates can vary from linear, branched, to completely compact and roughly spherical domain (Huang, *et al.* 2005). Further, these carbon aggregates may percolate and form agglomerates, which have the largest size (exceeding 1,000 nm) (Guerrero, *et al.* 2008). The easy formation of aggregate of the furnace carbon is due to the high inter-particular forces, while agglomerates result from weakly bound aggregates, and might be broken down into aggregates by micro-milling.

Properties of Carbon Black

Carbon black consists of aggregates, defined as the smallest dispersible units, which are composed of partially fused, reasonably spherical primary particles. Surface areas increase as primary particle size is reduced (Islam, *et al.* 2006). The aggregates, in turn, are held together by attractive van der Waals forces to form agglomerates. These forces increase as the size of the primary particle is reduced and agglomerate density is increased.

Characterisation of the surface properties of carbon black in terms of surface activity has been difficult compared to the other fundamental properties. However, surface activity has been evaluated in terms of the oxygen content and/or moisture adsorption rate (Khan, at al. 2010). For polar carbon black surfaces, oxygen content and acidity are important measures of the effective surface activity (Islam, *et al.* 2006).

Sulphur is one of the most common impurities in carbon black. It can be introduced with the tyre rubber feedstock. Inorganic contaminants in carbon black, such as Si, Al, and Cu can also be introduced with the rubber feedstock in the form of trace metals, cracking catalyst residues, and from the carbon black reactor (Sha, *et al.* 2006). It is necessary to obtain surface chemical states of these elements in terms of surface element distribution/concentration and chemical structure for the surface chemistry information on carbon black (Tscharnuter, at al. 2011).

Applications of Carbon Black

Traditionally, carbon black has been used as a reinforcing agent in tyres. Today, because of its unique properties, the uses of carbon black have expanded to include pigmentation, ultraviolet (UV) stabilisation and conductive agents in a variety of everyday and specialty high-performance products, including:

Tyres and Industrial Rubber Products:

Carbon black is added to rubber as a filler, and as a strengthening or reinforcing agent. For various types of tyres, it is used in inner liners, carcasses, sidewalls and treads utilising different types based on specific performance requirements.

Plastics:

Carbon blacks are now widely used for conductive packaging, films, fibres, mouldings, pipes and semi-conductive cable compounds in products such as refuse sacks, industrial bags, photographic containers, agriculture mulch film, stretch wrap and thermoplastic moulding applications for automotive, electrical/electronics, household appliances and blow-moulded containers.

Electrostatic Discharge (ESD) Compounds:

Carbon blacks are carefully designed to transform electrical characteristics from insulating to conductive in products such as electronic packaging, safety applications and automotive parts.

High Performance Coatings:

Carbon blacks provide pigmentation, conductivity and UV protection for a number of coating applications including automotive (primer basecoats and clear coats), marine, aerospace, decorative, wood, and industrial coatings.

Toners and Printing Inks:

Carbon blacks enhance formulations and deliver broad flexibility in meeting specific colour requirements.

Activated carbon:

Carbon black exhibits a high iodine absorption number making it a potential adsorbent for water purification and gold recovery.

Solid Fuel:

Carbon black has higher calorific value, fixed carbon and low sulphur and ash contents; it can be potentially used as high-quality fuel either in granules or briquettes.

METHOD

The process flow diagram shown in Figure 1 was developed for the pyrolysis of waste tyres in Gauteng Province, South Africa, as a benchmark to model the economic feasibility of this waste tyre treatment and utilisation option in the country. Gauteng, the smallest province in South Africa, occupying only 1.5 per cent of the land area, is highly urbanised and contributes up to 33.9 per cent of South Africa's Gross Domestic Product (GDP) valued at R811 billion. Gauteng consists of three main municipalities: City of Johannesburg (CoJ), Ekurhuleni (Ekh) and City of Tshwane (CoT) with a total of 48,077 tonnes per annum (tpa) of waste tyre generation. A total of three additional tyre pyrolysis plants, each with a capacity of treating 30 tonnes per day waste tyre/rubber material will be required to manage the waste tyres in Gauteng Province.

Process Description

As proposed in the REDISA plan, the waste tyres will be delivered by an approved waste tyre distributor to a permitted treatment facility where the total mass will be recorded at the weighbridge facility. The tyres' rings will be

Figure 1: Proposed flow diagram for tyre pyrolysis process

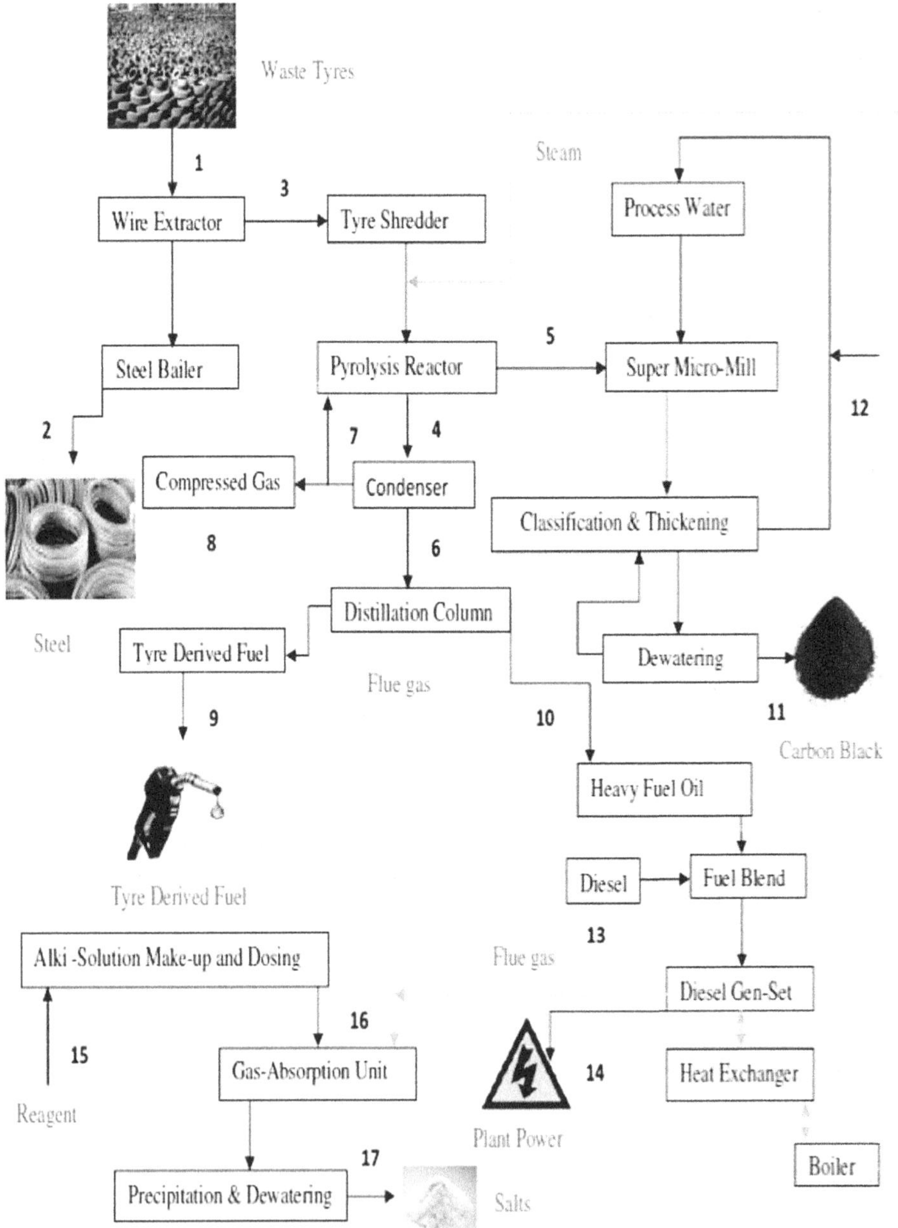

trimmed and cut into halves to extract the reinforced high tensile steel using a mechanical steel extractor. The steel cords will be baled and sold as a recyclable product. The trimmed tyres will be mechanically shredded to approximately 10–15mm rubber chips. The shredded rubber chips will be steam washed, dried and stockpiled before they are fed into a pyrolysis reactor vessel. The steam will be generated using excess heat from the diesel generator set as shown in Figure 1.

The reactor vessel will be initially heated using gas or fuel burners. This heats up the enclosed rubber chips in the absence of oxygen to temperatures of about 570°C until gasification occurs. The gases will be contained and condensed to form condensed heavy fuel oil (CHFO), which will be further fractionated in a distillation column to form light diesel equivalent tyre-derived fuel (TDF) and heavy fuel oil (HFO) (Sundarandian and Devaradjane, 2007). The HFO fraction will be blended with low sulphur commercial diesel fuel at a 30:70 volume ratio to fuel a diesel generator set that provides electrical power to all plant machinery.

This approach is considered the most economical option for combusting high sulphur HFO, since it allows for exhaust emissions capture and chemical neutralisation into value-added products. The emission from the generator set and reactor fuel burners will be contained and chemically treated through a gas absorption column to produce valuable products such as gypsum and sodium sulphite, depending on the type of alkali-solution used. The waste heat from the generator set and reactor cooling will be recovered though a heat exchanger for steam generation, which will be used in the plant.

Excess uncondensed gas from the process is recycled for process heating using gas burners, and a portion of this will be compressed and used as a fuel for forklift engines. At the end of each cycle, the reactor vessel is cooled; carbon black and steel remain as by-products. The steel is baled and sold as a recyclable product, whereas the crude carbon black will be further processed in a super micro-mill using process water to wash and facilitate wet grinding. The milled slurry product will be further classified and de-watered as a final product. Based on the waste tyre quantities in South Africa, a 30 tonnes per day plant is considered acceptable for conducting an economic and sustainability modelling study. A mass balance summary for the proposed plant is presented in Table 4.

Table 5 shows a costing summary of the proposed waste tyre pyrolysis plant capable of treating 30 metric tonnes per day of waste tyres. An

Table 4: Mass and energy balance for 30tpd waste tyre pyrolysis plant

#	Description	m_f (kg/h)	V_f (m^3/h)	Q_{out} (kW)	Q_{in} (kW)
1	Tyres	1,250	-	-	174
2	Steel	125	-	-	75
3	Tyre Chips	1,125	0.9	-	617.8
4	Pyrolysis Gas	667	834	68	68
5	Carbon Black	458	1.32	99	99
6	CTDO	563	0.63	-	-
7	Gas-Uncondensed	97	69	710	-
8	Gas-Compressed	8	6	57.2	-
9	TDF	375	0.465	-	-
10	HFO	188	0.22	-	-
11	Milled Carbon Black	458	-	-	367
12	Process Water	229	0.23	-	-
13	Diesel Fuel	79	0.09	-	-
14	Generator Set	267	0.31	1,100	-
15	Reagent Solution	98	-	-	15
16	Flue Gas	6,588	13,065	-	-
17	Sodium Sulphite	126	-	-	-

Table 5: Waste tyre treatment pyrolysis plant costing

Description	Cost ZAR
Steel wire extractor	R153,900
Tyre shredder	R974,700
Diesel Generator Set	R3,313,980
Fork Lifts	R180,000
Weigh Bridge	R212,840
Sub-Total Pre-Treatment	**R4,835,420**

Description	Cost ZAR
Hot Air Furnace c/w Heat Exchanger and Waste Heat Recovery Boiler	
Pyrolysis Reactor	
Gas Liquid Separation Tower c/w Flue Gas Treatment Tank	
Pumps and Tanks	
Interconnecting Pipes & Valves	
Control Unit	
Sub-Total Treatment	**R1,436,400**
Oil Distillation c/w Condenser	R1,846,800
Carbon Black Super Micro Mill	R1,128,600
Steel wire baler	R143,640
Slurry Handling Plant, Thickener and Filter Press	R17,442,000
Sub Total Post-Treatment	**R20,561,040**
Total Equipment	**R26,832,860**
Environmental Impact Assessment	R570,000
Fire Fighting Equipment	R280,000
Site Establishment	R419,300
Structural Design and Plant Enclosure Construction	R1,807,000
Professional Services	R853,601
Civil Design and Construction	R1,931,872
Electrical design and plant controls systems	R1,910,280
Installation and commissioning including supervision services	R1,824,975
Project Management	R588,000
Sub Total Engineering and Services	**R10,185,028**
Total Plant Cost Installed	R37,017,888
Critical, Commissioning & Operational Spares	R917,224
Shipping cost and import duties	R814,664
Contingencies	R1,731,888
Total Capital Requirement	**R42,451,565**

appropriate process was designed, taking into consideration the availability of waste tyres, environmental impact, product demand and market, and applicability of the technology in South Africa. All mechanical and process equipment was specified to the required process duties and its respective electrical requirements and civil loadings.

The technology offers modular pyrolysis reactors capable of treating up to 30 tonnes per day. Approximately 10 tonnes per cycle can be processed in a reactor and this will require one hour of loading, four hours processing and three hours of cooling and cleaning. Based on an eight-hour cycle, an average of eight tonnes per day can be processed. The lead-time for the supply of a complete pyrolysis plant is six weeks ex-factory upon confirmation of order. Shipping by sea is approximately six weeks and another 11 weeks will be required for installation and commissioning.

Utilities

The treatment plant will require process makeup water for the wet grinding of carbon black, cooling, washing and condensation of gases to crude heavy fuel oil. This water will be connected onsite via an allocated point to the municipal water supply or process water if available; the water consumption is approximately 0.23m^3/h. The pyrolysis reactors will be heated using pyrolysis gas at a consumption rate of 69m^3/h. Other components such as tyre cutters, shredders, mill drives, compressor plants, heaters, pumps, reactor drive units and control mechanisms require electrical power of 852 kW as shown in Table 6.

Table 6: Energy requirements for a 30tpd pyrolysis plant

Heating	677kW
Mechanical Energy	852kW
Cooling/Condensation	(366)kW
Plant Energy Requirements (combined)	1,530kW
Energy Efficiency	76%
Available Fuel Energy	767kW
Generator Set Output	900kW
Total Supply	1,666kW

Forklift operation will be run by compressed pyrolysis gas at an average consumption rate of 6m^3/h. The required plant power will be supplied by a generator set fuelled with a blend of diesel-HFO produced from the plant; excess thermal heat from this generator will be captured through heat exchangers and used for steam generation, which will be used to wash and dry the tyres prior to pyrolysis. Electrical power backup will be allowed for general plant lighting. There will also be a consumption of sodium hydroxide (NaOH) to treat the flue gas from the generator set and burner flues gases at 473 kg per day and at a cost of ZAR4,400/tonne. Sodium sulphite precipitate produced at 3 tonnes per day during flue gas scrubbing attracts a market price of about R209/tonne ex works.

RESULTS AND DISCUSSION

This economic model was based on a five-year project life, eight-hour working shift for three shifts in a day, and 25 working days per month as set out by the National Department of Labour. The model assumptions are summarised in Tables 7 and 8.

Table 7: Model Assumptions

Description	Value	Units
Annual Working Hours	8,760	Hrs/yr
Plant Estimated Downtime	1,560	Hrs/yr
Plant Available Time	7,200	Hrs/yr
Actual Plant Capacity	1.25	Tonne/hr
Exchange Rate	10.3	R/$
Project Period	5.0	Yrs
Depreciation Period	5.0	Yrs
Bank Lending Rate	6%	%/annum
Margin on Investment	3%	%/annum
Debt	100%	% Capital
Equity	0%	% Capital
Bank Finance Fee	2%	% on debt
Actual Annual Production	9,000	Tonnes/yr
Available Plant Capacity	10,950	Tonnes/yr
Actual Production	30	Tonnes/day

The plant cost was based on South African Rand at an exchange rate of R10.30 per US$1. A 100 per cent senior debt funding was used with a total interest rate of 9 per cent per annum to calculate the senior debt repayments taking into account company tax rate and value-added tax of 29 per cent and 14 per cent, respectively. A straight-line depreciation method was applied over a five-year period. Senior debt repayment of R11,896,235 was calculated over a five-year debt repayment term at 9 per cent per annum as shown in Table 8. An average inflation rate of 7.2 per cent per annum was used to calculate inflation indices of products, material and services. The costing also allowed for fixed and variable costs, including a 2 per cent per annum maintenance fee as a percentage of installed equipment cost, and a management fee as 5 per cent per annum of revenue as shown in Table 9.

Table 8: Input and output cost assumptions

Input Cost	Value	Units
Cost of Bulk Bags per Tonne	70	R/tonne
Electricity Cost	1.32	R/KWh
Diesel Fuel Cost	13.10	R/L
Water Cost	3.50	R/m^3
Cost of Sodium Hydroxide	4,400	R/tonne
Power Requirement	2,360	kWh/hr
Electrical Power Consumption	30	kWh/hr
Diesel Consumption	0.09	m^3/hr
Water Consumption	0.23	m^3/hr
Output Cost		
Sale of Sodium Sulphite	209	R/tonne
Sale of Tyre Derived Fuel	5.46	R/L
Sale of Refined Milled Carbon Black	950	R/tonne
Sale of Steel Wire Scrap Metal	680	R/tonne

The findings reported in this chapter are estimates from desktop studies. Detailed studies will be required for a more accurate costing exercise. Detailed cost calculations were conducted for a mobile plant of 30 tonne waste tyres per day production capacity with a total capital investment of R42,451,565. This amount includes equipment capital costs consisting of equipment as shown in Table 3, environmental impact assessments (EIAs),

Table 9: Financial summary of 30 tpd waste tyre pyrolysis plant

Term	Repaid	Interest	Outstanding
Advanced	R42,451,565		
0 yrs	-	R3,820,641	R46,272,206
1 yrs	(R11,896,235)	R4,164,499	R38,540,469
2 yrs	(R11,896,235)	R3,468,642	R30,112,876
3 yrs	(R11,896,235)	R2,710,159	R20,926,800
4 yrs	(R11,896,235)	R1,883,412	R10,913,977
5 yrs		R982,258R	R -

Record of Decision (ROD), site establishment and preparation; two years recommended and commissioning spares are allowed for. The expected revenue stream will be from variable end products such as refined carbon black, steel, and tyre-derived fuel and sodium sulphite. The operating costs as calculated in the model comprise labour requirements, input costs and plant equipment maintenance.

Table 10: Revenue summary of a 30 tpd pyrolysis plant

Period-Yrs	0	1	2	3
Sale of tyre-derived fuel	R5,462/m^3 **R5,899,928**	R5,856/m^3 **R19,593,316**	R6,277/m^3 **R21,004,035**	R6,729/m^3 **R22,516,325**
Sale of refined carbon black	R950/tonne **R1,252,746**	R1,018/tonne **R3,357,359**	R1,091/tonne **R3,599,089**	R1,170/tonne **R3,858,224**
Sale of sodium sulphite	R209/tonne **R76,119**	R224/tonne **R204,000**	R240/tonne **R218,688**	R257/tonne **R234,433**
Sale of steel wire scrap metal	R680/tonne **R244,800**	R728/tonne **R656,064**	R781/tonne **R703,301**	R837/tonne **R753,938**
Total Revenue	**R8,149,874**	**R25,623,171**	**R27,468,040**	**R29,445,738**

Four revenue streams were identified as tyre-derived fuel, refined carbon black, high-tensile steel and sodium sulphite. The recommended selling prices of the revenue streams presented in Table 10 above were set at a minimum and can be increased to enhance profitability. The variables include cost of diesel fuel at R5.46/L, electricity at R1.32/kWh, process water

Table 11: Fixed operational cost of 30 tpd pyrolysis plant

Period (Years)	0	1	2	3
Fixed Cost				
Plant Maintenance	R -	R910,162	R975,693	R1,045,943
2 x Supervisors	R325,000	R696,800	R746,970	R800,751
7 x Shift Operators	R455,000	R975,520	R1,045,757	R21,052
1 x Admin. Assistant	R127,400	R136,573	R146,406	R156,947
Cleaners 1-day shift	R58,500	R62,712	R67,227	R72,068
Total Salaries	R965,900	R1,871,605	R2,006,360	R2,150,818
Management fee	R407,494	R1,281,159	R1,373,402	R1,472,287
General Overhead	R1,282,705	R804,080	R861,974	R924,036
Auditing fees	R -	R19,296	R20,685	R2,175
1 x Security guard	R120,000	R128,640	R137,902	R147,831
IT	R -	R72,038	R77,225	R82,785
Insurance	R424,516	R455,081	R487,847	R522,972
Signage & Postage	R -	R4,288	R4,597	R4,928
Stationery	R3,000	R6,432	R6,895	R7,392
Legal fees	R22,000	R23,584	R25,282	R27,102
Furniture	R180,000	R -	R -	R -
Electricity fee	R120,000	R -	R -	R -
UIF & Skills Levy	R9,659	R18,716	R20,064	R21,508
Entertainment	R -	R4,502	R4,827	R5,174
Recruitment	R349,180	R -	R -	R -
Telephone	R6,000	R6,432	R6,895	R7,392
PPE	R36,000	R38,592	R41,371	R44,349
Total Fixed Cost	**R2,656,098**	**R4,867,005**	**R5,217,429**	**R5,593,084**

at R3.50/m^3, bulk bags at R70/tonne and sodium hydroxide at R4,400/tonne; and the fixed cost includes equipment maintenance, insurance, technical and operational staff as shown in Table 11.

Material and energy balances were conducted to quantify utilities and products on a yearly basis. These were linked to the input unit costs in order

Table 12: Variable operational cost of 30 tpd pyrolysis plant

Variable Cost	0	1	2	3
	R1.32/kWh	R1.42/kWh	R1.52/kWh	R1.63/kWh
Electricity	R9,112	R24,421	R26,180	R28,065
	R13.10/L	R14.04/L	R15.05/L	R16.14/L
Diesel	R3,525,249	R9,447,668	R10,127,900	R10,857,109
	R3.50/m³	R3.75/m³	R4.02/m³	R4.31/m³
Water	R2,308	R6,185	R6,630	R7,107
	R4,400/tonne	R4,716/tonne	R5,056/tonne	R5,420/tonne
Sodium Hydroxide	R249,279	R668,069	R716,170	R767,734
	R70/tonne	R75/tonne	R80/tonne	R86/tonne
Bulk Bag Cost	R92,308	R247,384	R265,196	R284,290
Total Variable Cost	R3,869,144	R10,369,306	R11,115,896	R11,916,240
Total Opex	R1,813/tonne	R1,693/tonne	R1,815/tonne	R1,945/tonne

Figure 2: Cash flow projections for a 30-tpd-pyrolysis plant

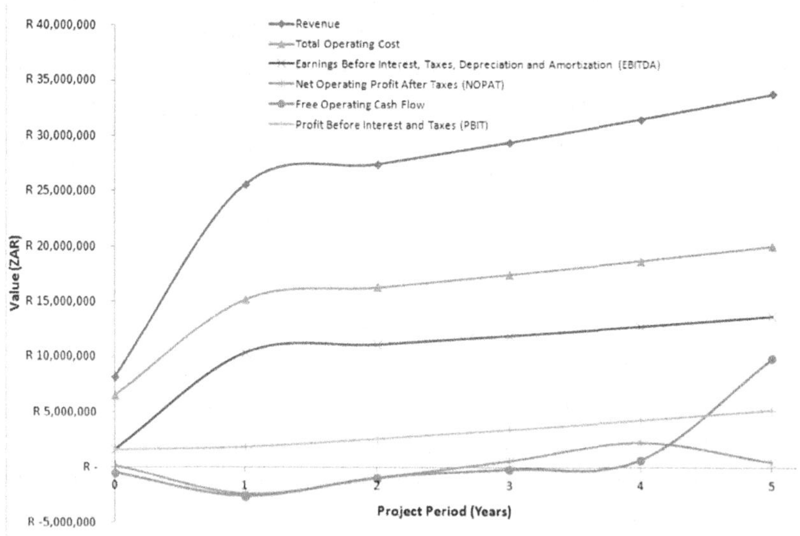

to calculate the actual treatment cost per unit mass of tyres, taking into account factors such as depreciation, debt repayments, and fixed and variable costs (Table 12). Figure 2 represents the cash flow projections over a five-year period, with revenue, total cost, and free operating cash flow. Based on this costing, the plant will yield a positive free operating cash flow of R685,741 between years three and four and the total treatment cost of R2,235/tonne of waste tyres.

The earnings before interest, taxes, depreciation and amortisation (EBITDA) of a project give an indication of the operational profitability of the business. It is defined by considering a project's earnings before interest payments. Tax, depreciation and amortisation are subtracted for any final accounting of its income and expenses (Suresh and Kazuteru, 2006). Depreciation is often a very good approximation of the capital expenditures required to maintain the asset base, so it has been argued that EBITA would be a better indicator. The gap between the project revenue and total operating cost represents the gross margin. A uniform increase on the revenue, EBITDA and operating cost (OPEX) is observed. This is influenced by the fact that Year 0 consisted of plant construction, installation and commissioning, resulting in reduced plant utilisation. A slight uniform annual increase in PBIT, EBITDA, OPEX and Revenues is linked to the 7.2 per cent annual inflation rate used in the model. A common trend between the free operating cash flow and net operating profit after tax (NOPAT) is observed during the first three years of the project. A positive free cash flow is realised in Year 3, which attracts income tax, resulting in reduced NOPAT. Between Years 4 and 5, there is a steep increase in free operating cash flow and a rapid decline in NOPAT as a large portion of the positive cash flow is exposed to income tax. It is expected to have a steady growing NOPAT and stable free operating cash flow immediately after Year 5 due to an end cycle of the senior debt repayment. Figure 2 shows a promising investment potential in waste tyre pyrolysis technology since the raw material is a waste material and can be procured at no cost. A gate fee is usually charged to the waste generator and this becomes an income stream for the waste processor. A debt-equity funding structure can also be considered whereby international technology suppliers and investors can provide a debt portion of the investment at an agreed return.

Approval of REDISA IWTMP by the national Minister of Environmental Affairs in terms of Regulation 6 (3) of the Waste Tyre Regulations in accordance to the National Environmental Management: Waste Act, 2008 (Act No. 59, 2008), allows REDISA to be in a position to collect up to R600 million per annum in the form of a tyre levy in order to support local entrepreneurs in waste-tyre recycling initiatives such as pyrolysis and crumbing with the aim of job creation, human capacity development and environmental management.

CONCLUSION

Waste tyre pyrolysis technology is one of the proven technologies that can be applied as an alternative method for the treatment and disposal of waste tyres and rubber products. Despite the findings discussed above, there is a need to assess the current waste tyre pyrolysis plants in South Africa in order to evaluate factors leading to the plants not being economically viable. Some of these factors include the following:

- There is a complicated process of obtaining environmental permits to operate such plants.
- There is a need for beneficiation and sale of carbon black as a supplementary revenue stream.
- None of South Africa's facilities is utilising excess gas produced in plant to optimise energy utilisation.
- Most technologies are sourced from China and do not comply with relevant SABS codes.
- Spare parts are not readily available in the country.
- There is no guarantee of feedstock (waste tyres) and product off-take at pre-agreed price and quantities.
- Most plants produce crude products of no value that require beneficiation. Therefore beneficiation of crude oil into valuable chemicals improves the plant process economics.
- There is a lack of knowledge of the technology and its relevance to the South African market.

The use of crude tyre-derived fuel as a direct fuel for any combustion system is not recommended due to the high level of contamination and high sulphur content. Refining of crude tyre-derived oil through a desulphurising distillation process is important in ensuring the production of clean low-cost fuel. Carbon black post-treatment via super micro-milling and classification helps to produce various grades of high-quality carbon black for the local and export markets. The heavy fuel fraction of the distillate can be used only with a low sulphur diesel blend in stationary combustion units such as generator sets and boilers fitted with emission neutralisation systems. The refined tyre-derived fuel is suitable for use in conventional diesel engines with blends of up to 1.5:1 volume ratio, tyre-derived fuel-diesel. Chemical treatment of flue gases from the combustion of heavy fuel oil results in the production of valuable sodium sulphite. The application of pyrolysis technology to the treatment of waste tyres was found to be economically feasible when considering a 30 tpd plant capacity with a required capital investment of R42.5 million with a potential investment return and gross margin of 27.1 per cent and 40.6 per cent, respectively. Each facility will produce an alternative 4 million litres per annum of refined tyre-derived fuel for application in diesel engines at the cost of R5.46/litre and other secondary value-add products for local and export markets. The use of alternative fuels by the municipalities will help in reducing fuel bills and dependency on conventional fuels such as diesel. This fuel can be used as a diesel additive by municipal-owned public transport such as buses, waste collection vehicles, and generator sets to run municipal wastewater works and pump stations.

This study supports the REDISA plan as it has the potential to make a significant contribution towards transport and waste social costs reduction, environment and climate protection, alternative sources of energy, green jobs creation, capacity building, skills transfer, postgraduate training, research and technological advancement in South Africa. There is no doubt that the availability of waste tyres and the product off-take markets are the key drivers for the successful operation of waste tyre pyrolysis technology.

References

Bajus, M. and Olahova, N. 2011. 'Thermal Conversion of Scrap Tyres', *Petroleum and Coal*, 53: 98–105.

Berrueco, C., Esperanza, E., Mastral, F. J., Ceamanos, J. and Garcia-Bacaicoa, P. 2005. 'Pyrolysis of Waste Tyres in an Atmospheric Static-Bed Batch Reactor: Analysis of the Gases Obtained', *Journal of Analytical and Applied Pyrolysis*, 74: 245–253.

Bhatt, P. M. and Patel, P. D. 2012. 'Suitability of Tyre Pyrolysis Oil (TPO) as an Alternative Fuel for Internal Combustion', *International Journal of Advanced Engineering Research and Studies*, 1: 61–65.

Bridgwater, A. V. 2003. 'Renewable fuels and chemical by thermal processing of biomass', *Chemical Engineering Journal*, 91: 87–102.

Cunliffe, A. M. and Williams, P. T. 1998. 'Composition of oils derived from the batch pyrolysis of tires', *Journal of Analytical and Applied Pyrolysis*, 44 (3): 131–152.

Department of Environmental Affairs. 2010. 'National Waste Management Strategy'.

ETRMA. 2011. *End of Life Tyres: A Valuable Resource with Growing Potential*, 2011 Edition. Brussels, Belgium: European Tyre and Rubber Manufacturers Association.

Fredonia Group. 2010. World Carbon Black Industry Study with Forecasts for 2013 and 2018, 1–8.

Gauteng Integrated Energy Strategy. 2011.

Guerrero, A., Goñi, S. and Allegro, V. R. 2008. 'Resistance of Class C Fly Ash Belite Cement to Simulated Sodium Sulphate Radioactive Liquid Waste Attack'. Eduardo Torroja Institute for Construction Science.

Hackett, C., Durbin, T. D., Welch, W., Pence, J., Williams, R. B., Salour, D., Jenkins, B. M. and Aldas, R. 2004. 'Evaluation of Conversion Technology Processes and Products for Municipal Solid Waste'. Integrated Waste Management Board Public Affairs Office, Publications Clearing House (MS-6).

Huang, K., Gao, Q., Tang, L., Zhu, Z. and Zhang, C. 2005. 'A Comparison of Surface Morphology and Chemistry of Pyrolytic Carbon Blacks with Commercial Carbon Blacks', *Journal of Power Technology*, 160: 190–193.

Islam, M., Parveen, H. and Haniu, H. 2010. 'Innovation in Pyrolysis Technology for Management of Scrap Tyre: a Solution of Energy and Environment', *International Journal of Energy Science and Development*, 1: 86–96.

Islam, M. R., Joardder, M. U. H., Hasan, S. M., Takai, K. and Haniu, H. 2011. 'Feasibility study for thermal treatment of solid tyres wastes in Bangladesh by using pyrolysis technology', *Waste Management*, 31: 2142–2149.

Jha, V. 2008. 'Carbon Black Filler Reinforcement of Elastomers', Ph.D. thesis, University of London, pp. 1–229.

Juma, M., Korenova, J., Markos, J., Annus, J. and Jelemesky, L. 2006. 'Pyrolysis and Combustion of Scrap Tyres', *Petroleum and Coal*, 48: 15–26.

Kalitko, U. 2008. 'Waste Tyre Pyrolysis Recycling with Steaming: Heat-Mass Balances & Engineering', 9: 213–236. Minsk: Belarus National Academy of Science.

Khan, N., Yahaya, E. and Faizal, M. 2010. 'Effect of Preparation Conditions of Activated Carbon Prepared from Rice Husk by CO_2 Activation for Removal of Cu (II) from Aqueous Solution', *International Journal of Engineering & Technology IJET-IJENS*, 10: 47–51.

Kiran, N., Ekinci, E. and Snape, C. E. 2000. 'Recycling of plastic waste via pyrolysis', *Resource Conservation and Recycling*, 29: 273–283.

Lehohla, P. 2012. 'Census in brief 2011'. Statistics South Africa, Pretoria Report Number 03-01-41.

Mahlangu, M. L. 2009. 'Waste Tyres Management Problems in South Africa and the Opportunities That Can be Created Through the Recycling Thereof', MA diss. in Environmental Management. University of South Africa.

Martinez, J. D., Puy, N., Murillo, R., Gracia, T., Navarro, M. V. and Mastral A. M. 2013. 'Waste Tyre Pyrolysis: A review', *Renewable and Sustainable Energy Reviews*, 23: 179–213.

Moghaddam, M. S., Moghaddam A. Z., Zanjani N. G. and Salimipour, E. 2011. 'Improvement fuel properties and emission reduction by use of Diglyme-Diesel fuel blend on a heavy-duty diesel engine'. 2nd International Conference on Environmental Engineering and Applications IPCBEE, Vol. 17.

Molewa, B. E. E. 2012. 'National Environmental Management: Waste Act (59/2008): National Waste Management Strategy', South African National Department of Environmental Affairs 563, 35306.

Molewa, B. E. E. 2013. 'National Environmental Management: Waste Act (59/2008): Approval of an Integrated Industry Waste Tyre Management Plan of the Recycling and Economic Development Initiative of South Africa', South African National Department of Environmental Affairs 569, 35927.

Morris, M. and Waldheim, L. 1998. 'Energy recovery from solid waste using advanced gasification technology', *Waste Management*, 18: 557–564.

Murugan, S., Ramaswamy, M. C. and Nagarajan, G. 2008. 'The use of Tyre Pyrolysis Oil in Diesel Engines', *Waste Management*, 28: 2743–2749.

Navarrete-guijosa, A. and Mostafa, S. T. 2007. 'A Comparative Study of the Adsorption Equilibrium of Progesterone by a Carbon Black and a Commercial Activated Carbon', *Journal of Applied Surface Science*, 253: 5274–5280.

Norman, D. T. 2008. 'Rubber grades of carbon black'. Product Development Witco Corporation, Concarb Division, Houston, Texas, pp. 1–19.

Nurulhuda, B. A. 2008. 'Preparation of Activated Carbons from Waste Tyres Char Impregnated with Potassium Hydroxide and Carbon Dioxide Gasification', MSc thesis, pp. 1–24.

Pilusa, J. and Muzenda, E. 2013. 'Qualitative Analysis of Waste Rubber-Derived Oil as an Alternative Diesel Additive', *International Conference on Chemical and Environmental Engineering (ICCEE)*.

Pilusa, J., Muzenda, E. and Shukla, M. 2013. 'Molecular Filtration of Rubber Derived Fuel', International Association of Engineers.

Qu, W., Zhou, Q., Wang, Y., Zhang, J., Lan, W., Wu,Y., Yang, J. and Wang, D. 2006. 'Pyrolysis of Waste Tyre on ZSM-5 Zeolite with Enhanced Catalytic Activities', *Polymer Degradation and Stability*, 91: 2389–2395.

Quek, A. and Balasubramanian, R. 2013. 'Liquefaction of Waste Tyres by Pyrolysis for Oil and Chemicals – A Review', *Journal of Analytical and Applied Pyrolysis*, 101: 1–16.

REDISA, 2012. 'Recycling and Economic Development Initiative of South Africa. Waste Tyre Volumes per Province'.

REDISA, 2013. 'Recycling and Economic Development Initiative of South Africa. Waste Tyre Volumes per Province'.

Schrenk, H. H. and Berger, L. B. 1941. 'Composition of Diesel Engine Exhaust Gas', *American Journal of Public Health and the Nations Health*, Vol. 31 No. 7: 31, 669–681.

Shah, J., Rasul, J.M., Mabood, F. and Shahid, M. 2006. 'Conversion of Waste Tyres into Carbon Black and their Utilization as Adsorbent', *Journal of the Chinese Chemical Society*, 53: 1085–1089.

Sienkiewicz, M., Kucinska-Lipka, J., Janik, H. and Balas, A. 2012. 'Progress in used tyres management in the European Union: a review', *Waste Management*, 32: 1742–1751.

Steffens, D. 2003. 'The Diesel Engine and the Environment'. Dynamic Positioning Committee, Marine Technology Society Environment. Wartsila North America, Inc.

Sundarandian, S. and Devaradjane, G. 2007. 'Performance and Emission Analysis of Biodiesel Operated CI Engine', *Journal of Engineering, Computing and Architecture*, 1(2): 1–22.

Suresh, B., and Kazuteru Y. 2006. CEH Marketing Research Report (Zurich: Chemical Economic Handbook SRI Consulting), pp. 771.1000A–771.1002.

Tawil E. 2008. 'Boiler Fuels, Emissions and Efficiency – Course No: M02-028'. Continuing Education and Development, Inc.

Tscharnuter, W. W., Zu, L. and Weiner, L. 2011. 'ASTM Carbon Black Reference Materials: Particle Sizing Using a Brookhaven Instruments BI-DCP, Disc Centrifuge Photosedimentometer Theory'. Brookhaven Instruments Corporation, 1: 1–10.

Weil, E. D. and Sandler, R. 1999. Kirk-Othmer *Concise Encyclopaedia of Chemical Technology*, 4th edition, p. 1937. New York: John Wiley & Sons, Inc.

Williams, P. T. 2013. 'Pyrolysis of waste tyres: A review'.

CHAPTER 9

Energy-efficient Low-income Housing Development in South Africa: The Next Build Programme

Lyndall (Lynda) Mujakachi

ABSTRACT

South Africa's pathway to a low-carbon Green Economy is established in various frameworks and policies. These various policies promote the following, inter alia: 1) identifying opportunities for new areas of growth and economic participation; 2) setting the country on a new growth and development path; and 3) job creation and poverty reduction.

Since 1994, South Africa has been building houses for the low-income wage group, that is, those earning R3,500 per month or less. A very small number of these houses have incorporated green energy measures, in spite of the fact that there are policies that have been developed since 1994 that promote sustainable development and the need to move to a low-carbon economy. The low-income housing development is a sector for which government has a captive market and there is opportunity for government to push for the deployment of proven green energy technologies on a large scale and hence create new industries.

However, the move from policy to implementation seems to present a challenge: the measures that need to be taken have been articulated in various policies and the technology is available. The mechanisms to bridge policy to action are not well coordinated. Implementing the green energy strategies will require strong political commitments, administrative actions and financial investments to effect the bold decisions.

INTRODUCTION

Electricity and housing (or human settlements) development have been identified by the ANC government as sectors that can contribute to the socio-economic transformation of South Africa (RDP[61] 1994: 31, 22). These sectors were identified as contributing to the improvement of livelihoods and advancing the marginalised Black population into mainstream economic activities, as well as galvanising new manufacturing activities in the economy.

The South African Constitution stipulates that its citizens have the right to access adequate housing. The Reconstruction and Development Programme framework endorsed the right to housing. Since 1994, the government has embarked on a national programme to provide housing for the low-income wage earners (people earning R3,500 per month or less) and the indigent[62] in urban and rural[63] areas. Furthermore, the RDP framework pronounced the provision of electricity to include the Black majority – '*electricity for all*.' The government has made great strides in the provision of electricity for Blacks in urban and rural areas. In 1994, 36 per cent (comprising mostly the white population) had access to the 'cheap and abundant' electricity (RDP 1994). It is estimated that over 80 per cent of the Black population in urban and rural areas now have access to electricity.

In terms of meeting basic needs of the South African population, the RDP (1994) framework announced the following housing standards: provide protection from weather; a durable structure; reasonable living space and privacy; sanitary facilities; storm water drainage; a household energy supply (linked to grid electricity supply or other sources such as solar energy); access to clean water (RDP 1994: 23). With reference to energy development, the

61. Other sectors identified in the RDP 1994 are: water and sanitation; telecommunications; transport.

62. In identifying indigent people, indicators such as type of housing are used, e.g. informal settlements, public housing/hostels, backyard shacks, RDP housing; and/or social groups, e.g. disabled people dependent on the state, children, the unemployed, and the elderly who depend on a pension or state grant (van Reyneveld, *et al.* 2003).

63. According to the White Paper on Local Government 1998, the definition of 'urban' and 'rural' areas is based on a loose categorisation of settlement types. Under 'urban', there are three sub-categories: (i) urban core, characterised by high population density with over 10 dwelling units per hectare, high levels of economic activities and high land values; (ii) urban fringe: these are settlements on the boundaries of municipalities, but outside the urban core, and include low-income settlements and high-income low-density settlements; and (iii) small towns: these vary greatly, but most are economically and socially linked to surrounding rural hinterlands. 'Rural' areas are also listed according to sub-categories and have populations of around 5,000 or less inhabitants.

RDP emphasised energy efficiency and conservation. The following were pronounced: improvement of dwelling thermal performance; promotion of energy-efficient appliances; use of solar water heaters; implementation of time of use electricity tariffs; utilising sustainable energy sources such as solar power (RDP 1994: 31–32). The provision of renewable energy technologies and energy-efficient housing has been promulgated in various policies and legislation; for example, the White Paper on Energy Policy in South Africa (1998); White Paper on Renewable Energy (2003); Energy Efficiency Strategy (2005); South African National Standards (SANS 204); New Growth Path (2011); Integrated Resource Plan (2010); White Paper on a Climate Change Response Strategy (2012); Infrastructure Development Act (2014).

Since 1994, the government has built some 3 million[64] low-income houses (or RDP houses in the local parlance). It is further estimated that some 2.3 million[65] people do not have housing, meaning they do not have their own house and/or are living in an informal settlement (Presence, 2014). The figure is rising owing to the rapid rate of urbanisation and, according to the Minister of Human Settlements, Lindiwe Sisulu, the housing delivery programme has fallen by an average of 25 per cent over the past five years due to 'blockages in the pipeline' (Presence, 2014).

A small number of the newly built houses have incorporated green energy technologies and these have been mainly pilot retrofitted housing units. The provision of housing is more than just brick and mortar. Low-income housing development should be targeted with a developmental and sustainability motive, as pronounced in the various frameworks and legislation, and it is about empowerment of individuals and communities. More often than not, the ownership of a house is the single biggest investment that a low-income earner will make in his or her lifetime, and this has the potential to improve the welfare of the low-income and indigent groups. The low-income housing development is a sector in which government has a captive market and there is opportunity for government to push for the deployment of proven green energy technologies on a large scale and hence create new markets and industries.

The aim of this chapter is to review the development and implementation of programmes to provide energy-efficient housing to low-income and indigent households. These programmes are being implemented by municipalities and Eskom (in areas that a municipality does not distribute electricity) and facilitated by the Department of Energy (DoE). The

64. These are estimates. Lack of monitoring makes it difficult to establish accurate figures.
65. The figure ranges from 2–3 million. There has been no proper monitoring of low-income houses built since 1994.

Department of Human Settlements' (DHS) mandate is to provide sustainable housing and, hence, it has to collaborate with these and other agencies. As stipulated in the various policies and legislation, energy-efficient low-income housing development can contribute to setting the country on a new growth trajectory through the emergence of new industries from the development of new technologies, job creation and poverty reduction, and achieve social equity. This chapter contributes to the ongoing debate on implementing energy-efficient low-income housing and proposes some recommendations on the way forward.

The chapter starts with a conceptual framework; the next section reviews various programmes that are being implemented with regard to availing electricity to low-income and indigent households and the role that renewable electricity is playing; followed by a section that presents a case study of a greenfield development comprising energy-efficient low-income housing; and the last two sections deal with the way forward and the conclusion.

CONCEPTUAL ISSUES

South Africa has adopted the principle of sustainable development to guide socio-economic development. Sustainable development is set to place the country on a developmental trajectory towards greater efficiency and innovation in resource use and the integration of social, economic, ecological and governance systems (NFSD, 2008). Green economic development is part of the continuum to achieve sustainable development. Green economic development is defined as:

> *a system of economic activities related to the production, distribution and consumption of goods and services that result in improved human well-being over the long term, while not exposing future generations to significant environmental risks or ecological scarcities. It implies the decoupling of resource use and environmental impacts from economic growth* (Green Economy Summit Report 2011: 4).

Energy efficiency, in the context of sustainable development, refers to using energy whilst ensuring that the environmental impacts of energy conversion and use are minimised (DoE, 2005: 1). South Africa's economy is largely based on minerals extraction and processing, which is energy intensive. In the past, South Africa has taken for granted its 'cheap and abundant' electricity, but this changed when, in the 2000s, demand began to outstrip supply. Energy-efficient low-income housing development, as articulated in the previous section, includes: improvement of dwelling thermal performance; promotion of energy-efficient appliances; a household energy supply from the grid and other sources such as solar energy; and a durable structure (RDP, 1994).

Energy-efficient low-income housing development presents an interesting case study in terms of transitioning to renewable electricity. The building of low-income houses does not encompass the larger vision of 'development' and 'sustainability', even though these terms are invoked to justify that low-income housing provision contributes to social and human development, and ultimately economic development. The low-income housing programme is characterised by piecemeal measures: achieving target numbers of housing units built; implementing energy efficiency features such as pre-paid meters; and providing solar technology for houses that are not connected to the grid. The goal of transitioning to clean and renewable energy sources, as pronounced in various policies, seems to be lost in energy-efficient low-income housing development, and yet, this is an area where government has a captive market.

As part of South Africa's international commitments to reduce carbon emissions,[66] its national objectives to achieve sustainability,[67] and as part of the programme to provide electricity to marginalised South Africans, there have been programmes implemented by government to provide renewable energy technologies for electricity generation to households that are not on the grid in urban and rural areas.

A review of the programmes being implemented shows disparate programmes. Energy-efficient low-income housing development can be a catalyst to transition to renewable energy technologies. Until the introduction of the Renewable Energy Independent Power Producer Procurement Programme (REIPPP) for the supply of electricity into the grid, there was not much motivation, especially from Eskom, the national utility, to move towards renewable energy, as the technologies were considered expensive compared to the cost of coal-powered electricity supplied by Eskom. The electricity power cuts that were experienced in 2008,

66. South Africa ratified the UNFCCC in August 1997 and acceded to the Kyoto Protocol in July 2002. About 95 per cent of electricity generation is from coal-fired power stations.

67. In addition to the various frameworks and legislation, also refer to the *Long-term Mitigation Strategy* (LTMS) for South Africa (DEA 2011).

and the subsequent rapid rise in electricity prices,[68] have seen government include renewable energy as part of the substantive energy supply mix (Integrated Resource Plan 2011) and its courting of the private sector to contribute to electricity generation from renewable energy sources.

This chapter adopts the 'energy transition approach' (Lilliestam, *et al.* 2012) as a conceptual framework (Figure 1) to analyse managing energy transition. Energy transition is a systemic or transformative change in energy systems. In the context of sustainable development, energy transition refers to renewable energy.[69]

Figure 1: Energy transition framework

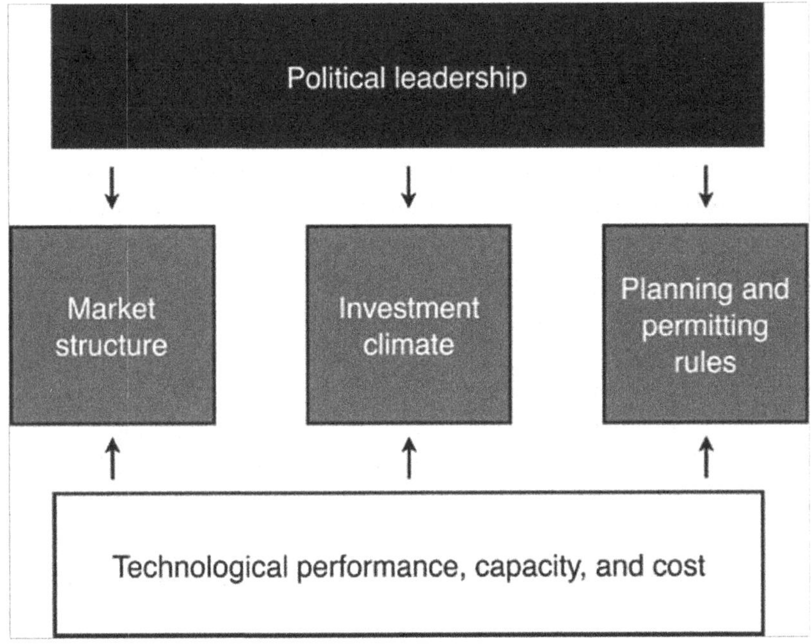

Source: Lilliestam, *et al.* 2012

The following briefly describes the energy transition framework that has been adapted and contextualised to the energy-efficient low-income housing programme in South Africa. The political leadership that operates from the top down involves government institutional infrastructure (that is, various government departments and quasi-government organisations in the three

68. Eskom, the national electricity utility, has been increasing prices annually to finance the building of two new coal-fired power stations, Medupi and Kusile, to augment electricity supply.

69. Renewable energy technologies are: (i) solar [CSP and PV]; (ii) wind; (iii) biomass; (iv) ocean waves; (v) hydro power; and (vi) waste.

spheres of government[70]) and institutional frameworks (policies, laws and planning systems); technological performance operates from the bottom up and this involves research organisations and universities that carry out the research in technology development, as well some private sector[71] companies. The three systems in the middle are where changes need to occur. These three are: the markets, that is, the low-income housing sector where power is bought and consumed; new investment in renewables and infrastructure, mostly by the private sector and government; and the planning and permitting rules that dictate the pace and placement of local economic development which should take place at municipal level (Lilliestam, *et al.* (2012: 2). It should be noted that this framework is a representation of the real world and will go through iterative changes during implementation. Transitioning to renewable energy for electricity generation requires groundwork, not only in experimenting with new technologies to bring their costs down, but in experimenting at sufficient scale so as to force the institutional development issues that come with them (Lilliestam, *et al.* 2012: 1).

POLICY IMPLEMENTATION IN ENERGY-EFFICIENT LOW-INCOME HOUSING

The government has enacted various policies and legislation that guide electricity development in post-1994 South Africa so as to bring access and equity of electricity provision for all citizens, particularly the marginalised Black (African, Coloured and Indian) people. The key policies are briefly described. The White Paper on Energy (1998) announced the post-1994 government policy on transforming the energy sector and emphasises, *inter alia*, facilitating electricity access to Black people. The White Paper on Renewable Energy (2003) pronounced the promotion and implementation of renewable energy as part of the energy supply mix that includes coal. The government set a target of 10,000GWh renewable energy contribution to final energy consumption by 2013 (this is yet to be achieved). The Energy Efficiency Strategy (2005) presents guidelines for the implementation of energy efficiency practices as mandated in the White Paper on Energy (1998) and links energy sector development with socio-economic development plans.

Ruiters (2009) noted that the bulk of the electricity generated in South Africa is consumed by industry and commerce. It is estimated that the

70. The three spheres are national, provincial and municipal governments.
71. Some large corporations carry out their own in-house technology development research.

residential sector consumes less than 20 per cent of electricity generated and the low-income households consume much less. Households pay more on electricity tariffs compared to industry and commerce, that is, R0.29 per kWh and R0.16 per kWh respectively. The affordability factor is important to secure energy in households, especially in low-income households (2009: 249).

What are some of the programmes that have been developed and are being implemented in terms of ensuring access, affordability and security of electricity for low-income and indigent households? In answering this question, an attempt is made to investigate the impetus to the implementation of renewable electricity. The section will look at the National Electrification Programme, free basic electricity (FBE), the pre-paid electricity meter system, free basic alternative energy (FBAE), the solar water heating (SWH) subsidy programme, Solar Home System (SHS), and the Renewable Energy Independent Power Producer Procurement Programme (REIPPP).

National Electrification Programme

The National Electrification Programme was the post-1994 government's response to electrify urban low-income and rural households – announced as '*electricity for all*' in the RDP (1994) framework. The programme was implemented by Eskom and municipalities and funded by government. The justification for this national programme was to improve the welfare of low-income and indigent households and stimulate some income-generating activities. The programme stipulated that both grid and non-grid power sources such as solar energy would be deployed.

In the mid-1990s (pre- and post-apartheid era), significant levels of electrification in urban areas were achieved. In the post-2000 period, the government and Eskom focused on availing electricity to rural areas. It has been noted that the South African electrification programme is considered an outstanding achievement (in terms of infrastructure development) by a government of a developing country, producing results that have not been duplicated by other developing countries (Mohlakoana, 2014). A notable development was how Eskom, an apartheid-era organisation, was able to rapidly transform itself and fit into the new political dispensation. Eskom became proactive in considering political concerns during and after the political transition period by aggressively supplying electricity to Black households in urban and rural areas (Mohlakoana, 2014; Marquard, 2006).

According to Mohlakoana 2014:

> *The implementation of this electrification policy was also made possible by the availability of generating capacity and the "creative (low-cost) ways" (that is, using single-phase instead of a three-phase connection for low-income areas households as they often use less electricity due to lack of and insufficient incomes) that were used by Eskom to electrify low-income rural areas where most of the households had no electricity* ... [and quoting Matinga 2010] ... *the South African electrification program has differed from electrification programs in other Sub-Saharan African countries, because the country has largely financed its own program* (2014: 33).

Subsequent programmes (see below) implemented in urban and rural areas under the auspices of the National Electrification Programme have resulted in over 80 per cent of the population having access to energy. The low-income households in urban areas were supplied with electricity through connection to the grid (hence, it was mainly electricity from Eskom's coal powered stations). Many Black urban residential areas and some rural areas (that were close to the grid) benefited from the programme. This programme set the foundation for embarking on the national electrification agenda.

Free Basic Electricity

The publication of the White Paper on Energy Policy in 1998 emphasised that low-income households were to have access to adequate basic energy services for cooking, heating, lighting and radio and television, regardless of capacity to pay. The ANC government announced the provision of free basic electricity supply for low-income households in 2000 in its election manifesto (Ruiters, 2009; Mapako and Prasad, 2005). Implementation began in 2003 and is being facilitated by municipalities that distribute electricity and Eskom. Funds for this programme are allocated to municipalities through the Equitable Share grant disbursed by the Department of Cooperative Governance and Traditional Affairs (DoE, 2007). No funding is provided to Eskom from national government but Eskom claims for FBE from municipalities in terms of a service level agreement between Eskom and each municipality.

Free basic electricity amounts to 50kWh[72] per month. This provides basic lighting, basic media access, basic water heating using a kettle, basic ironing to a low-income household that is connected to grid electricity (Ruiters, 2009: 251) The DoE stipulates that it is not the intention of government to provide 'free electricity', but 'free basic electricity' and government has deemed 50kWh per month as sufficient electricity to meet the needs of a low-income household as these households generally have a low demand for electricity (DoE). The amount of electricity that is supplied as free basic electricity was based on the recommendations of research conducted for Eskom and the DoE by the Energy Research Centre (ERC) at the University of Cape Town between 2001 and 2002 (Dugard, 2009: 273).

Additional electricity requirements for indoor heating, hot water geysers and cooling systems are charged at the approved tariff rate and this expenditure has to come out of the low-income household budget. The reality is that a low-income house (for example, comprising a three-person family using refrigerator, heater, stove, television, and hot water geyser) consumes on average 498kWh and can be expected to pay between R200 and R300 per month (Ruiters, 2009: 250) and the price of electricity has been rising since 2009. The free basic electricity allocation of 50kWh per month is definitely an inadequate electricity supply. Some low-income households supplement their FBE allocation by using firewood (which is mostly collected for free) and paraffin for cooking, and candles for lighting.

According to Ruiters (2009):

> *A developmental approach to the level of supply might take a different view of the minimalist understanding of needs. It is self-evident that low consumption is a reflection of the apartheid legacy, unemployment and a produced inability to consume. To change this pattern of unhealthy under-consumption, more than 50 kWh would have to be offered. In fact, municipalities often use 150 kWh as their cut-off for deciding whether or not households should be seen as poor* (2009: 251).

While South Africa prided itself on supplying cheap and abundant electricity prior to 1994, which was accessed mostly by the white population, the provision of electricity to the larger population post-1994 also necessitated the increase in generating capacity to meet rising demand. This, as well as the growth of the economy, has seen electricity tariffs escalating rapidly,

72. Municipalities that can afford to subsidise more than 50 kWh per month are encouraged to do so.

especially since 2008, as Eskom embarked on a programme to build new power stations (Medupi and Kusile). Most low-income households cannot afford the increasing electricity tariffs and electricity supply is frequently being cut off due to non-payment.

Pre-paid Meter System

In an effort to address affordability of electricity by low-income households, a system of pre-paid meters was introduced. South Africa is a recognised leader in the manufacture and installation of pre-paid meter equipment and this has developed into a lucrative industry. Black Economic Empowerment (BEE) deals from this new industry benefited those who were politically connected (Ruiters 2009: 257). The justification for introducing pre-paid meters were, *inter alia*, each household would be able to manage its electricity consumption; less administration in terms of recovery of debts, less disconnections and reconnections; and upfront payment improves municipal cash flow (Ruiters 2009: 257). With the introduction of pre-paid meters, some jobs were lost, notably those of meter readers. It should be noted that, while pre-paid meters were initially introduced as a measure to control unaffordable electricity consumption in the low-income residential areas, pre-paid meters have found their way into middle- and high-income[73] households. In middle- and high-income residential areas, the voluntary adoption of pre-paid meters in households is to manage and control electricity expenditure. In the low-income residential areas, pre-paid meters were a tool to control access to electricity to households that are not able to pay, thus effectively limiting adequate access to electricity and thus being unable to realise the '*electricity for all*' mantra.

The demand for electricity in low-income residential areas has resulted in most of the households not being able to afford adequate electricity, resulting in their electricity supply being cut off for non-payment. The latter has given rise to illegal connections – '*Izinyoka*' – (electrocution accidents are on the increase leading to injuries and fatalities) (Annegarn, *et al.* 2010) and public protests demanding 'affordable electricity' are becoming commonplace. In 2012, Eskom was owed R2 billion in electricity arrears by some residents in Soweto.

73. A salary of above R15,001 per month is considered middle- to high-income. This figure is based on figures used to allocate low-income housing by government. Hence, a low-income household earns R3,500 or less per month. A 'gap market' is where a household earns R3,501 to R15,000 but does not qualify for a bank loan to buy or build a house, hence the government steps in to assist.

Free Basic Alternative Energy (FBAE)

The government provides free basic alternative energy (FBAE) in recognition of indigent households in urban and rural areas that are not connected to the electricity grid (DoE, 2007). The programme is administered by municipalities and funded through the Equitable Share Grant, which is disbursed to municipalities by the Department of Cooperative Governance and Traditional Affairs. The FBAE policy stipulates that, where no electricity infrastructure exists, the funding for FBE (see above section) 'must be channelled to fund FBAE and municipalities are encouraged to supplement the FBE grant from their own income [thus] ensuring that indigent households receive FBAE' (DoE, 2007, Chapter 2: 2.1).

The criteria for identifying areas that are to be provided with FBAE to indigent households are: most distant from the electricity grid; where no Solar Home System[74] (SHS) Programme is planned; where there are no immediate plans to electrify the area; and/or where energy poverty is prevalent (DoE 2007). The energy carriers that are being funded under the programme are paraffin, liquefied petroleum gas (LPG), coal, bio-ethanol gel, candles, firewood and biogas. FBAE is allocated to the value of R55 per month and the policy stipulates that the figure should increase by the inflation rate plus 1.5 per cent (DoE, 2007). In implementing the FBAE programme, a municipality may use its own personnel or external service providers (DoE, 2007).

The implementation of FBAE has not been without challenges. Firstly, in many municipalities there is a lack of indigent policy[75] and no registration of indigents. Secondly, the FBAE policy stipulates that municipalities are required to supplement the FBAE allocation from their own budgets (DoE, 2007). Municipalities have continuously indicated that they are not allocated the annual budgets they request and hence are operating with tight budgets. Thirdly, most communities are not aware that there is provision to access FBAE. Fourthly, the provision of FBAE is more expensive than FBE since most of the alternative energies are not regulated.

Lastly, lack of affordable and safe alternative energy is an aggravating factor in a low-income household. In a study conducted by the Paraffin Safety Association Southern Africa (PASASA) in informal settlements, there was evidence that the majority of households used paraffin for cooking, heating and lighting (PASASA, 2012: 8) as this was the energy carrier the said households could afford. The study noted:

74. See below for an analysis on Solar Home System (SHS).
75. Refer to footnote 63 above.

> *Currently, only one locally manufactured, non-pressurized, licensed, legal stove exists which is sold under the brand name Panda. Although this product is reported to be used by 80% of households in South Africa, most often, households are in possession of an unlicensed, unsafe version of the Panda which is being manufactured by an unknown source and dominating the market due to its low-cost, wide distribution network and ease of use. This unlicensed "old" Panda does not comply with South African standards and regulations as outlined by the South African Bureau of Standards (SABS) and the National Regulator for Compulsory Specifications (NRCS), however, efforts to trace and prosecute these manufacturers have, as yet, been unsuccessful. The structure of these micro-enterprises and the distribution of fuel and appliances more widely have created a situation in which compliance with national standards and regulations is difficult to assess and enforce* (Pasasa, 2012: 8).

'Shack' fires caused by faulty appliances are commonplace in informal settlements. These fires do not affect one household only but spread to several shacks in the informal settlement as there are no fireguards between the shacks. The main victims of shack fires are children due to lack of adequate care and supervision (PASASA, 2012: 9). Furthermore, some low-income households and informal settlements that do not have electricity connections are resorting to illegal connections. This results in injuries and fatalities from electrocution (Annegarn, *et al.* 2010).

In all, energy carriers such as paraffin, coal and firewood, which are used by most low-income households, have health, safety and environmental concerns. Coal and firewood contribute to poor indoor and outdoor air quality and this can lead to respiratory problems (Annegarn, *et al.* 2011). Low-income households have aspirations to access modern sources of energy such as grid electricity (preferred) and/or solar power systems. The rapid increases in price of electricity are further delaying the replacement of combustion sources as a key source of energy in low-income households.

Solar Water Heating (SWH) Programme

The solar water heating (SWH) subsidy programme was introduced by Eskom in 2008. The SWH programme was one of Eskom's demand-side management strategies in response to the electricity load shedding that took place in 2008. Solar water heaters have two designs: high-pressure systems and low-pressure systems. High-pressure systems are being installed in medium- and high-income households and the low-pressure systems are being installed in low-income households as part of the government's retrofit programme. Eskom affords a partial subsidy for the installation of high-pressure systems, which are mostly fitted in medium- to high-income households and there is a full subsidy for low-income households where low-pressure systems are mostly installed. The SWH programme has witnessed the nascent emergence of small and medium enterprises (SMEs) involved mainly in importing and selling equipment. Much of the SWH equipment for the low-income household retrofit programme was imported from China, as it was less costly (Ledger, 2011).

In 2012, Eskom and the Department of Trade and Industry (dti) issued a government directive that low-pressure SWH equipment, which was being installed as part of government's retrofit programme, was to have 70 per cent local content of components manufactured. Many businesses that had won tenders to install solar water heaters thrived from importing the equipment, mostly from China. When Eskom invited new tenders for the low-cost systems in 2012, none of the companies complied with the local content rules (Bega, 2015).

This may seem like a knee-jerk decision on the part of Eskom. However, as a state-owned enterprise (SOE), Eskom has a responsibility to ensure that local manufacturing capacity is developed. This necessitates the restructuring of procurement regulations. For example, the government has a goal to install 4 million solar water heaters in homes by 2030. There should be a phased process of procurement: in the short-term, some equipment may need to be imported while local manufacturing capacity is being developed (this requires industry to work in tandem with research organisations, that is, universities and the Council for Scientific and Industrial Research (CSIR); based on monitoring and evaluation information, decisions can be made on phased gradation thresholds that should be achieved in local content manufacturing by 2030. So, let us say that a company responds to a request for a tender in supplying solar water heaters in year 10 of the programme (assuming 2015 is the base year), and the threshold for local content is then pegged at 70 percent; the company cannot claim 'previous disadvantage' or any other limitation because the regulatory instrument (assuming it was promulgated in year 2015) would have stipulated the trajectory of local content requirements. The onus is on prospective companies to embark on a pathway to develop the requisite capacity. This entails that government should have a programme that assists companies, especially SMEs, in innovation and technology development and deployment. The Innovation Hubs that are being established are designed to offer such assistance. Furthermore, this entails that government may need to develop a computer register of low-income housing development targets as public information over a three- to five-year rolling cycle and such information should be updated on a regular basis.

Eskom discontinued the SWH subsidy programme in December 2012. The SWH programme was transferred to, and will be administered by, the Department of Energy. In February 2015, the Department of Energy issued a media release on the way forward in implementing the SWH programme. One of the conditions is that the new rebate system will introduce subsidies on a sliding scale commensurate with local content and products with the highest verified local content will be awarded the highest rebate (DoE, 2015).

Solar Home System (SHS)

Non-grid electrification through concessionaires was launched in 2001. The Solar Home System (SHS) was identified as a solution to provide electricity to rural households that were not connected to the grid. The low load demand, the dispersed nature of rural settlements, the high fixed costs of grid extensions, and the ever-rising costs of electricity made grid electrification more and more uneconomical and unsustainable in the long term, resulting in most rural areas not having access to electricity. The imperative to adopt modern energy systems in rural areas was to address human and biophysical health and safety issues, and alleviate risks associated with the use of wood, paraffin and candles. The call to reduce carbon emissions made off-grid electrification attractive (DoE, 2012).

There are six private sector service providers (or concessionaires) who are providing services in the designated concession areas. The rights last for five years, but the service contract remains binding for a period of 20 years (DoE, 2012). The current concessions are in the following areas:

No.	Concessionaire	Province
1.	KwaZulu Energy Services (KES)	KwaZulu-Natal and Eastern Cape
2.	Nuon RAPS Utility (Pty) Ltd	KwaZulu-Natal
3.	Solar Vision (Pty) Ltd	Limpopo
4.	Ilitha Cooperative	Eastern Cape

No.	Concessionaire	Province
5.	Summer Sun Trading (Pty) Ltd	Eastern Cape
6.	Shine the Way cc	Eastern Cape
7.	KES (German (KfW) donor funded)	Eastern Cape

Source: Non-Grid Household Electrification Guidelines (DoE, 2012)

The basic SHS comprises: a photovoltaic (PV) panel (50Wp) (this equates to about 250Wh per day); a charger controller; wiring and outlets for small appliances; a battery (105AMP-hour) and four energy-efficient compact fluorescent lights (CFLs) (DoE, 2012). The SHS must meet the government specification for PV use in households. With SHS, rural households can watch television for up to four hours, have lighting for four hours, use a portable radio for 10 hours, and charge cellphones. Rural households continue to use firewood, coal, paraffin, and LPG gas for cooking, heating and cooling to supplement their energy requirements. The Department of Energy has made it a requirement that concessionaires of SHS should augment their services by selling thermal fuels.

The financing model that is being used is the fee-for-service model. The initial investment is borne by the concessionaire, thus eliminating high upfront investment costs by poor rural households as the main source of income for these households is from the government grant system. The onus is on the concessionaire to provide a service that is affordable to rural households. The DoE subsidises 80 per cent of the capital cost and the concessionaire pays the remaining 20 per cent. Customers pay the concessionaire a once-off connection fee of R100. A monthly service fee of R58 is charged to the customer to cover lifetime running costs including the operation, maintenance, replacement of batteries, fee collection, customer service, and so on. The maintenance of the system is borne by the concessionaire (DoE, 2012). The rural households receive a subsidy of R40 per month, which is paid by the respective municipality to the service provider (Mapako and Prasad, 2005). The subsidy is from the free basic electricity (FBE) programme. Hence, rural households pay R18 per month for SHS service.

In a study carried out in the Eastern Cape, it was observed that the poorest households were excluded from benefiting from FBE due to the selection criteria employed to qualify for installation of an SHS. Poorer households could not afford the initial once-off connection fee of R100 and the monthly service fee of R58 for an SHS. Further, it was noted that SHS was not supplying the electricity needs of a household; hence, households had to supplement their energy needs with wood and paraffin (Mapako and Prasad, 2005). Most rural households depend on the government grant as their source of income. The said study further noted:

> *The absence of the subsidy at the commencement of SHS dissemination in the rural areas meant that the solar off-grid electrified poor rural customers paid a fixed monthly fee of R58. The number of SHSs on offer was restricted by the amount of subsidy the government could pay to the service providers/concessionaires. In Eastern Cape Province, by the end of 2003, some two years after installation started, and some six months after the launch of the FBE benefits, the subsidy was still not implemented, prejudicing poorer households further. The service provider had removed systems from households unable to pay the monthly R58 regularly* (Mapako and Prasad, 2005).

Renewable Energy Independent Power Producer Procurement Programme (REIPPP)

The Integrated Resource Plan (IRP 2010, promulgated 2011) has set out the amount of electricity that South Africa will require by 2030 and the various technologies that will be employed to generate electricity. The IRP stipulates that the plan will be continuously revised and updated as necessitated by changing circumstances (IRP, 2011: 7). The IRP heralds a new age of energy generation in that renewable energy has been allocated the following amounts: solar photovoltaic (PV) 8,400MW; concentrated solar power (CSP) 1,000MW; wind 8,400MW; and hydro[76] power 2,609MW. Nuclear[77] energy was allocated 9,600MW. Gas was allocated 6,280MW.

Coal will continue to contribute substantially to the energy mix with 10,133MW committed and new build options amounting to 6,250MW (IRP 2010 updated 2013: 12). There is an ongoing research project on carbon capture and storage (CCS) that is testing the feasibility of capturing carbon dioxide and storing it underground. The CCS project is being conducted by the South African National Energy Development Institute (SANEDI) together with mining and industrial companies (*Financial Mail*, 2014: 10); however, CCS is a technology that is still a long way from commercial realisation (Eltrop, 2011).

76. Hydroelectric power will be imported and there will be local small-scale generating plants.

77. Nuclear power is from the beneficiation of uranium. While nuclear power is carbon-free, it faces waste disposal problems (Foxon, 2002).

The transition to renewable energy generation for electricity is being realised through the Renewable Energy Independent Power Producer Procurement Programme (REIPPP), which operates under the auspices of the IRP 2010. While the REIPPP may not have been specifically designed to address electricity requirements in low-income and indigent households, some valuable lessons will cascade to the SHS and SWH programmes in terms of technology development and local manufacturing.

The REIPPP has opened the pathway to providing renewable electricity generation and augmenting Eskom's electricity generating capacity from its coal-fired power stations. The REIPPP will, in the main, benefit industries and urban areas as it is has been designed to feed into the grid. It should be noted that when the building of the two new power stations, Medupi and Kusile, is completed, some of Eskom's current fleet of power stations will need to be decommissioned as they will have reached the end of their lifespan and new electricity generation capacity will be required. Renewable energy generation for electricity is expected to have an increasing role.

In the bids that have taken place under the REIPPP, private sector developers have been awarded 20-year contracts for the development of solar, wind, mini-hydro, biomass and landfill gas projects to supply electricity to the national grid. To date, a total of 79 projects have been commissioned under the REIPPP since 2011, representing combined generating capacity of 5,243MW that will be supplied into the grid. The 5,243MW is just above the output from one of Eskom's new coal-fired plants, Medupi or Kusile (each will generate 4,800 megawatts). These are projects that have mobilised some R168 billion of private sector funding (Lund, 2015). The REIPPP is instrumental in bringing a decline in prices of renewable energy supply to the national grid through the competitive bidding process. It remains to be seen how the falling delivery costs of renewable energy for electricity will translate into a saving and benefit for the consumers, especially the low-income households.

The introduction of REIPPP for the supply of electricity into the grid has begun the process of introducing new technology and the development of new industries. As part of the contractual obligations under the REIPPP, developers have to create jobs and local manufacturing and community upliftment (*Financial Mail*, 2014: 21). This is a welcome development as South Africa has been progressively experiencing de-industrialisation and jobs are being lost in areas which employ semi- and unskilled persons, that is, agriculture, mining and manufacturing. In Ekurhuleni, which is the main manufacturing hub of South Africa, there is a lot of industrial floor space lying idle due to the decline in manufacturing. This can be used to push-start the green economy through the manufacturing of renewable electricity components and building materials. The metals and engineering industries, which were the cornerstone of the apartheid economy, have been stagnating since 2000. The dti is envisaging local content in manufacturing to increase to a 45 per cent threshold by round three of the REIPPP bidding process and eventually reach 65 per cent.

Under the REIPPP, a process of distributed spatial development is emerging. The various projects that are being implemented in renewable electricity generation can be found throughout the country: solar projects can be found in Eastern Cape, Western Cape and Northern Cape Provinces; wind is being established in North-West Province and solar PV is being established in the Free State (Ntuli, 2014, pers. comm.).

Energy-efficient Low-income Housing

The Department of Energy (DoE) developed guidelines – Energy Efficient Strategy (2005) – for, *inter alia*, the construction of thermally designed housing incorporating passive solar design. The Energy Efficient Strategy (2005) is supported by the South African National Standards (SANS) 204 Energy Efficiency in Buildings (2011). The SANS regulation makes it compulsory for all new buildings to be designed and built to standards that minimise the energy required to meet functional requirements. The Energy Efficiency Strategy (2005) further announced the replacement of electric geysers with solar water heaters (SWH).

Energy efficiency measures comprise passive features such as: north-facing orientation; insulation in ceilings; roof overhang; window sizes that accommodate maximum lighting and ventilation; construction materials for floors and walls that resist water absorption and provide thermal insulation (that is, warm in winter and cool in the warm to hot seasons), as well as introducing renewable energy technologies into households. The SANS regulation provides a detailed and comprehensive description of the building standards required, including water saving features, energy-efficient bulbs, building requirements and standards for each climatic zone (South Africa has six climatic zones) (SANS, 2011).

With an estimated 2.5 million backlog in low-income housing development, this presents a huge potential market for the development and deployment of new technologies such as energy-efficient building materials and renewable energy equipment. The existing low-income houses are another potential market to deploy energy-efficient technologies through retrofitting. The DoE, in collaboration with the Department of Trade and Industry (dti), has presented a trajectory of local manufacturing content thresholds with regard to the REIPPP and SWH programmes (see above).

The current subsidies allocated for low-income housing construction do not incorporate renewable energy technologies. However, passive energy efficiency features can be afforded by the current subsidy. It is incumbent upon the National Department of Human Settlements to revise the housing subsidies in light of the national mandate to achieve energy security and energy efficiency and to re-industrialise the economy. For example, the subsidy in the Western Cape province has been increased from R75,000 to R110,000, which covers the construction of a 40 square meter house with two bedrooms, a bathroom and combined living and kitchen area, and passive energy efficiency measures (Verwey, 2014), and this should be further increased to include renewable electricity technology to provide adequate supply for lighting, cooking, refrigeration, communication (television and radio and charging cellphones) and heating. Pilot projects that are incorporating the installation of SWH, solar panels, wind energy, and water tanks in low-income houses have been funded from donor sources (Verwey, 2014). It is important that government explores new types of funding instruments and the amount of funding required that include renewable energy technologies in low-income households.

With regard to retrofitting, it has been observed that many low-income houses are poorly built and are structurally unable to accommodate installations such as rooftop solar panels and solar water heating. The media (print and electronic) has on several occasions publicised cases of housing units in which walls are cracking before they are occupied. It is important that building inspectors monitor and verify that contractors adhere to building standards.

The above interventions demonstrate that the South African government has been implementing programmes to facilitate and ensure electricity provision in low-income and indigent households. Various policies have been promulgated indicating that there is political will to introduce renewable energy generation for electricity.

The FBE programme is being impacted by the rapidly rising cost of electricity, and the 50kWh is insufficient to meet household energy requirements for lighting, cooking and heating. Low-income households that are not connected to the grid and/or cannot afford to pay for additional household electricity requirements have to resort to alternative energy sources such as coal, paraffin and firewood (these are energy sources that are provided in the FBAE programme), and these have attendant safety, health and environmental concerns. The demand for electricity has witnessed increasing illegal activities associated with the need to access electricity, that is, illegal extensions and connections to houses that are not connected with electricity, and illegal reconnections and tampering of meters (pre-paid meters) once the amount of electricity units purchased has been used (Ruiters, 2009: 259).

The projects on implementing SHS and SWH indicate that renewable energy technologies are available, but they have to be implemented at sufficient scale to create market demand. Furthermore, the prices of solar PVs have been gradually reducing. Energy-efficient low-income housing presents an opportunity to implement renewable energy technology.

However, there is need for institutional capacity building at municipal level to facilitate the development of decentralised renewable energy as part of local economic development. The programmes that are being implemented in various municipalities do not seem to monitor lessons learnt from the various projects and what interventions can be scaled up to the wider national level, like the electrification programme undertaken by Eskom and municipalities in the 1990s and 2000s. These government initiatives have the potential to yield to market demand. The challenge is for the three spheres of government to collaborate and develop capacity to coordinate the various projects, to develop a monitoring and evaluation system, and to channel the 'lessons learnt' information into policy and into a national development programme. The Department of Performance Monitoring and Evaluation in the Presidency is mandated to lead this process.

The introduction of renewable energy technologies and the implementation of passive energy-efficient low-income housing development herald a new developmental trajectory. South Africa's success in achieving electrification to some 80 per cent of low-income and indigent households in urban and rural areas risks being negated by the escalating price of electricity, and there is a push-back to the use of combustion fuels (that is, wood, coal and paraffin). In the energy sector, government has been striving to reduce energy poverty, improve health and safety, provide affordable and adequate electricity to low-income and indigent households, and job creation.

ENERGY-EFFICIENT LOW-INCOME HOUSING DEVELOPMENT: WITSAND IN CAPE TOWN[78]

The following section presents a case study on a 'greenfield' integrated housing development project that incorporates energy efficiency features. While the introduction of energy efficiency and renewable technologies in low-income houses is through subsidised retrofitting, the Witsand project demonstrates a development project that includes energy efficiency features from the design stage. This is also a landmark project in that it incorporates and consolidates various legislations that mandate the provision of energy-efficient low-income housing.

The Witsand project in Atlantis, Cape Town is an upgrade of a previous informal settlement. This area had no infrastructure and was inhabited by approximately 1,000 informal dwellings. Sustainable human settlements development in the Witsand context aimed at providing an integrated shelter with energy and water saving features, urban agriculture and empowerment (skills transfer and job creation) to the local people (Guy, 2010).

PEER Africa, a not-for-profit organisation, has been implementing the Witsand low-income housing project in partnership with the City of Cape Town. PEER Africa was also collaborating with the University of Johannesburg through the EnerKey project that was conducting research on renewable energy development, as well as the University of Cape Town with regard to monitoring and evaluation. PEER Africa has interpreted various government policies and developed an integrated human settlement model which has been branded as iEEECO™ (integrated Energy Environment Empowerment and Cost Optimisation) for the low-income groups.

PEER Africa began engagements with the City of Cape Town and the Witsand community in 2000. There was a delay in project approval due to changes in housing policy. Further delays were experienced in getting approval for housing development, land acquisition and the bureaucracy of subsidy applications, which led to the need for extensive conflict resolution as residents grew increasingly dissatisfied with the pace of delivery.

The original site plan that had been drawn up by the City of Cape Town was redesigned in 2002 by the PEER Africa project team assisted by town planning (MCA Town Planners) and engineering (ASCH Consulting Engineers) consulting companies. The site plan was designed as an integrated/mixed residential development including social and economic

78. The information for this case study is based on various interactions with PEER Africa, that is, making joint presentations at various municipalities in Gauteng; joint meetings with various Gauteng provincial government officials; presentations by PEER Africa and the University of Johannesburg at various conferences and workshops; a study tour undertaken by the author to Witsand in November 2010 and project reports from PEER Africa.

amenities and passive solar layout design. The site design was completed with the participation of the community housing committee. Workshops were held with the community in the informal settlement so that the residents could contribute to the proposed development project.

The project was approved for the Phase 1 development (passive solar site plan, facilitation of the community housing subsidy beneficiaries and self-help construction of 400 housing units) in June of 2004 and the flow of government funds started in October 2004. The Phase 1 pilot was completed in 2006 and the National Energy Regulator of South Africa (NERSA) and the University of Cape Town (UCT) certified the approach that had been adopted by PEER Africa. PEER Africa intended to carry forward the lessons learned to Phase 2 and was targeting a scaled development of 1,000 passive energy-efficient low-income houses and the installation of some renewable energy technologies. The City of Cape Town issued a new request for proposal between May to June 2009 and PEER Africa submitted its proposal.

In the initial contract with the City of Cape Town, PEER Africa was contracted as a sole source service provider in terms of the Municipal Finance Management Act (MFMA) as PEER Africa was assisting the municipality in implementing an innovative human settlement project in which they had the expertise. According to the MFMA, a service provider can only be contracted for three years. Since technology roll-out has to be tested over a period of time (a minimum of at least 10 years) in order to realise economic and social benefits, the three-year stipulation in the MFMA is an impediment for public-private partnerships, especially partnerships that introduce new ideas. The Phase 2 residential construction phase of the project was targeted to start in the last half of 2010, but the project encountered delays. In the meantime, PEER Africa has continued to engage with the community and the City of Cape Town and carries out monitoring of the project benefits.

The settlement layout is designed to maximise the use of free energy from the sun, that is, dwellings are north facing and incorporate thermal-efficient building materials. It should be noted that at first the residents were apprehensive about living in houses that did not face the street and/or were facing the back of the neighbour's house. In low-income housing development, the infrastructure design is done separately from the top structures and this creates a number of challenges for architects and developers to achieve passive solar and other free basic energy benefits once the construction work starts.

The province and the City of Cape Town approved an additional 52 units to be constructed during Phase 1 that saw the size of the housing units increase from 36 square metres to 40 square metres. A number of additional features were included in the new designs including: upgraded roof structures that included various types of roofs, e.g. tile roofs, enhanced window designs and double storey units (densification). Some housing units were built according to the further increased size of 50 square metres per unit. The project piloted the following renewable technology features: solar water heating, solar home system (LED lighting and cell charger), water savings and rooftop storm water recovery.

The Witsand housing project developed a spiral hob that runs off-grid. This product can be used for cooking on the solar heating system. Some private companies have developed a small solar home system that includes a cellphone charger and lighting that can go into dwellings.

Since 2010, Peer Africa has been sourcing for international aid funding to continue the work in Witsand. The project has received international recognition and accolades and PEER Africa staff and Witsand community members are invited to make presentations at conferences. PEER Africa has continued to engage with the City of Cape Town and the Witsand community.

Besides energy efficiency benefits, there are health and safety benefits that go beyond the normal benefits of the low-income housing programme, such as reduction in indoor pollution and amelioration of fires due to the elimination of open flame for space heating (Guy, 2010). With specific reference to coastal areas, there is strict adherence to coastal area building specifications leading to the reduction in mould formulation due to coastal condensation (Guy, 2010). The Witsand low-income housing project provided important lessons for the municipality in drafting procurement and tender documents in terms of how a municipality can incorporate new ideas that are required for improved service delivery.

Overall, the project has demonstrated a low-income housing development model that offers reduction in household energy costs, water demand management, and reduction in household pollution. The next phases of the project would have wanted to include household solid waste recycling, storm water recovery and reuse, as well as wind energy and greening measures.

While the Witsand housing project has been recognised as a groundbreaking development, there has been no formal adoption by the City of Cape Town. Some of the residents who have benefited from the iEEECO™ housing development have presented at various workshops and have called for the project to be scaled up. Moreover, various provincial and municipal government officials from Gauteng have conducted study tours to Witsand. It is, however, difficult to ascertain how new ideas are interrogated, internalised and adopted (or thrown out) in municipal government. The imperative to implement innovative programmes such as energy-efficient low-income houses that contribute to service delivery and the potential to create local economic development is lost in government inertia.

WAY FORWARD

The low-income housing sector presents a ready market to introduce new renewable technologies and other innovative features. The estimated 3 million low-income houses[79] that have been built since 1994 can be retrofitted with renewable electricity technologies and some municipalities have embarked on pilot projects. The estimated housing backlog of 2.5 million presents an opportunity for adopting integrated greenfield developments as pronounced in various legislation. The low-income housing market is a captive market for the deployment of new sustainable housing materials and renewable electricity technologies.

With reference to the energy transition framework, this section will reflect on the role of government and innovation and market formation to assist in moving forward energy-efficient low-income housing development.

Roles and Responsibilities of National, Provincial and Municipal Governments

The roles of national, provincial and municipal governments are identified in the Constitution. This section will concentrate on the evolving roles as per the National Development Plan, and focus on the municipal government as it is the sphere of government that is at the front line of implementing development activities.

Since 1994, South Africa has implemented public sector reforms that are to ensure that the three spheres of government operate as a seamless entity and not in silos. In a developmental state, towards which the ANC aspires, the national government acts as the centre that sets policy frameworks. The provincial government, over and above provincial mandates as defined in the

79. The Department of Human Settlement has to ensure monitoring of the housing delivery programme.

Constitution, should define the strategies and coordinate municipal activities as per provincial jurisdiction. The municipalities are the level of government that, together with the communities, are meant to act as change agents in local communities. There should be a coordinated mechanism of monitoring and evaluation that has a bottom-up approach and feeds into the policy process. Furthermore, innovation hubs are being established that can play an important role in facilitating linkages between municipalities and with R&D organisations.

The government published the National Development Plan (NDP, 2011) that sets out a developmental vision until 2030 and identifies target areas, for example: development of a low-carbon economy (renewable electricity falls under this target area), sustainable human settlements and other economic infrastructure. It is incumbent upon national government to provide the oversight and coordination in terms of policy. The provincial and municipal governments should concentrate on implementation, monitoring and evaluation. For example, South Africa is a signatory to the United Nations Framework Convention on Climate Change (UNFCCC). Before the national government had developed a national climate change policy, Western Cape and Gauteng Provinces developed respective provincial strategies. The City of Johannesburg developed its climate change policy before Gauteng Province. It should be recognised that provinces and municipalities have plenty to do without being caught in a perennial cycle of policymaking. If one examines the actions that need to be implemented to address climate change, they are already existing policies: e.g. energy efficiency, renewable energy development and sustainable housing policies.

The term 'developmental local government' was coined in the Local Government White Paper in 1998 whereby local government is committed to working with citizens within the community to find sustainable ways to meet their social, economic and material needs and improve the quality of their lives (Atkinson, 2002: 9). Before 1998, municipalities were primarily concerned with the operations and maintenance of basic functions within their localities. Since the publication of the Local Government White Paper (1998), a new developmental role for municipalities has become paramount. This involves a much more strategic, innovative and multi-sectoral involvement in infrastructural, economic, and social development (Atkinson, 2002: 26).

The provincial and local governments have been in a cycle of policymaking and strategy-making as regards their developmental role since

2000. The practical steps and interventions required to operationalise policies should be enunciated in the Integrated Development Plan (IDP). The IDP is a legal document as per the Municipal Systems Act of 2000. The IDP interprets macro-visions (policies and strategies from national and provincial government) and translates and consolidates the same into 'municipal reality' (action plans and budgets). The process of drafting IDPs in many municipalities, which should be an in-house activity, has been mainly out-sourced to consultants, who use the previous plans as a basis for the subsequent IDP. Hence, much of the IDP continues with its previous momentum and small incremental steps are made annually. The IDP should include green economy projects as announced in various policies (Annegarn, 2013). Hence, skills development, the process of internalising plan crafting through continuous practice (learn-by-doing) and making strategic choices for development, should be a focus area in terms of municipal capacity building.

Currently, the major source of revenue in municipalities is from the distribution of electricity, which is bought from Eskom and on-sold to municipal residents. In order to decouple local government revenue generation from electricity consumption, officials need to consider alternative revenue streams. Local government is required to play an increasing role in facilitating local economic development.

Let us look at some factors that need to be addressed to effect the transition to energy-efficient low-income housing development.

Distributed Renewable Energy Generation

Schedule 4 of the Constitution states low-income housing development is a provincial competence area (albeit, in recent times, large metros have been delegated to provide low-income housing), and electricity (and gas) reticulation is a municipal competence area. In practice, both Eskom and municipalities distribute electricity to domestic consumers. Eskom's domestic customers are mostly urban low-income households and rural households that were electrified as part of the national electrification programme. It is important for national government to provide guidelines, in consultation with provinces and municipalities, on distributed electricity development, that is, electricity generated from households with rooftop solar panels and feeding excess capacity into the grid.[80] In addition, the technology for solar PV systems to use batteries to store electricity for use after sunset or days when there is no adequate sunlight is rapidly evolving so that solar PV generators can deliver utility-scale power on demand (Chown, 2014).

Public Private Partnerships to Facilitate Innovation

A regulatory instrument needs to be developed to facilitate public-private partnerships[81] for introducing innovations into municipalities. The current three-year tenure for a sole sourcing contract under the MFMA is inadequate for new technology roll-out. Technology has to be tested over a period of time through an iterative process of monitoring and evaluation in order to realise economic and social benefits. Such a partnership between the municipality and an external partner (whether it be a private company or university or research organisation) requires an effective contractual arrangement so that both parties win. The service provider should be able to get a return on investment. The municipality has the regulatory and oversight role but should attract new investments to reduce service delivery backlogs (Atkinson, 2002). However, many councillors are uncomfortable with external organisations that may be deemed to benefit from providing services to the municipality, especially since there is the political imperative of Black Economic Empowerment (BEE) which seeks to spread job creation and wealth creation to the larger Black community. Furthermore, most councillors operate on the basis of their elected terms. The Infrastructure Development Act (2014) supports long-term infrastructure development that goes beyond a single electoral tenure. This necessitates the updating of the MFMA to meet the requirements of the Infrastructure Development Act (2014).

Procurement Regulations

Procurement regulations for service providers should stipulate adherence to existing regulations. For example, SANS 204 has comprehensive guidelines for the construction of energy-efficient houses in the various climatic zones, but these are not being implemented. Low-income housing development has continued to perpetuate apartheid-era standards and/or maintain the status quo, that is, basic housing that is connected to the Eskom grid for electricity supply. Progressive contracts in low-income housing retrofitting programmes should be designed to motivate and incentivise local manufacturing capacity of renewable electricity equipment. The

80. This refers to residential areas that are near the grid.

81. *Investigate Municipal Infrastructure Investment Unit (MIIU) to facilitate Municipal Services Partnerships (MSPs)* (ref. Atkinson, 2002).

announcement by the DoE that the new SWH rebate system will introduce subsidies on a sliding scale commensurate with local content and products with the highest verified local content will be awarded the highest rebate is a welcome development (DoE, 2015).

There are concerns over the promotion and introduction of technology in low-income housing developments without a clear contractual understanding of the service level required and the detailed legal service level contract. The Non-Grid Household Electrification Guidelines (DoE, 2012) provide information such as application process by municipalities, funding, contractual agreements, maintenance, monitoring and reporting, re-use of de-installed equipment and community awareness and education. These guidelines will have to be scaled up in future guidelines from the current minimalist understanding of low-income household requirements and take into account a more holistic and developmental approach to the provision of non-grid or renewable energy for electricity generation.

Funding for Energy-efficient Low-income Housing Development

The current low-income housing subsidies can accommodate passive energy efficiency features but do not include renewable energy equipment installations at the construction stage. It is important that the subsequent low-income housing build programme should include installation of appliances such as solar PV panels, SWH and so on at the construction stage. The inclusion of renewable energy equipment will contribute to the growth and expansion of the manufacturing industry in South Africa. The development of a sustainable market (with potential to export) will see a decrease in the cost of the equipment. The Department of Human Settlements (DHS), DoE and National Treasury will need to devise innovative funding mechanisms for scaling up the inclusion of renewable energy technologies.

A further motivation to accelerate energy-efficient housing development that includes renewable energy technologies is that, currently, Eskom is responsible for providing the infrastructure and installing electricity. Eskom regulations stipulate that the utility will not provide electricity until 80 per cent of the construction work on low-income houses at a project site has been completed and the houses are occupied. As most, if not all, municipalities are experiencing high demand for housing, most low-income housing recipients move into accommodation that is without electricity.

When the renewable energy technology becomes fully developed, households will be able to sell their excess electricity to Eskom by feeding into the grid. This is what is happening in Germany and other European countries that are implementing distributed renewable electricity generation.

Future funding should incorporate sustainable design criteria and installation of appropriate technologies that include the following: thermally efficient design, sustainable building materials, energy efficiency, renewable energy options, sustainable water and sanitation systems, and waste minimisation and recycling.

Municipal Government Officials are Change Agents

A developmental municipality is required to facilitate new ideas that will improve service delivery. Political and management buy-in is critical in ensuring that new ideas are institutionalised and implemented. One of the key roles of a municipality is to be a change agent. Some municipal officials have embedded ideas and resist change. For example, PEER Africa had to negotiate to change the existing site plan and design so that it could accommodate the north-facing orientation to achieve passive energy. This was initially met with resistance from town planners who wanted to maintain the 'tried, tested and proven' process of site planning for low-income residential areas where houses face the street.

Mohlakoana (2014) concluded that policy implementation processes do not only fail due to lack of financial resources. Municipal officials, as change agents, should demonstrate the following characteristics: motivation, cognition (internalising and interpreting new ideas and information), and the capacity and power to influence the policy implementation process (2014: 337). Capacity building of municipal officials and councillors, which includes study tours to learn from practical experiences in other countries, is an ongoing programme. However, there is concern that there are no tangible outcomes to the municipal capacity building programme, given the challenges of service delivery.

Annegarn (2013) posited the following:

> *It is important that all levels of government should be made responsible to deliver on the green economy in a coordinated manner. The move to a low carbon economy has to be a key performance indicator (KPI) of officials at all levels* (Annegarn, 2013).

Monitoring and Evaluation

Monitoring and evaluation is an important function in service delivery programmes of government. The purpose of monitoring and evaluation is to monitor and evaluate the results of the implementation, provide a basis for decision-making regarding the enhancement of existing programmes, promote accountability, and ensure that best practices are documented and inform policy and new developments.

The various policies and Acts stipulate that monitoring and evaluation has to be embedded in all programmes. However, the lack of enforcement capacity within government is a significant limiting factor and this needs to be addressed with capacity building of municipal officials. For example, the amount of FBE (50kWh) per month is still based on a study conducted in 2002. There is a need to assess the impact of this programme on low-income households and to better appreciate the changing needs of these households.

Innovation and Market Formation

South Africa has the capacity to implement innovative programmes on a national scale; for example, the national electrification programme and the Motor Industry Development Programme (MIDP), which demonstrate innovation and market formation. The major drawback in instituting energy efficiency and renewable energy technologies in low-income housing development was the low price of electricity prior to 2008 compared to the cost of generation using renewable options. The rapidly rising electricity prices since then, the falling prices of green technologies, as well as load shedding has provided an impetus for the implementation of renewable energy technologies.

There are various projects that the government is implementing with regard to deploying renewable energy technologies in low-income houses; for example, the implementation of solar home systems and solar water heaters. However, there does not seem to be a process of developing these projects beyond the current small-scale initiatives. Further, it is important to monitor the programme and document the lessons that emanate from implementing the various projects that address energy efficiency and the introduction of renewable technologies in low-income housing. As such, they can be evaluated, important lessons consolidated, and scaled up to a national infrastructure development programme.

The low-income housing market presents a unique opportunity to deploy renewable energy technologies. The onus is on municipal governments to internalise, translate and champion national policies for implementation.

CONCLUSION

This chapter has presented some of the green energy development activities that are taking place in the low-income housing sector as pronounced in various policies. While the said projects demonstrate that some implementation is going on, this remains at pilot and/or small-scale levels. The lack of monitoring and evaluation of most projects has resulted in the absence of information on lessons learnt for development of future projects and programmes.

South Africa has the capacity to implement large-scale national programmes. The electrification programme has demonstrated wide-ranging impact on the welfare of low-income and indigent households. The national utility, Eskom, further demonstrated a capacity and innovative ideas to ensure that the goal of *electricity for all* is largely achieved via the national electrification programme. However, more can be done to move towards renewable electricity development, especially in low-income and indigent households. While innovative pilot projects are being implemented to address access and affordability to the wider population, there is a need to have a developmental approach that looks at adequate free basic services per household.

A policy framework exists to enable the national implementation of energy-efficient and renewable energy technology in the development of low-income houses. National and provincial government departments have to coordinate their work, as some of the enabling legislation resides in different spheres and departments. Municipalities have to incorporate green energy development in their respective IDPs as pronounced in various policies.

Research organisations and universities need to be more intensely involved in the green energy development programme. The policymakers require information that is based on sound research so that they make informed decisions. The implementation of a national project on energy efficiency and renewable energy technology deployment for low-income housing has to be accompanied by continual monitoring and evaluation and this capacity has to be developed in all municipalities.

The energy-efficient low-income housing and renewable energy development have the potential to transform the economy through innovation and the development of new industries. Energy-efficient low-income housing development is an area where government has a captive market.

References and Further Reading

African National Congress. 1994. *The Reconstruction and Development Programme – A policy framework*. Johannesburg: Umanyano Publications, ISBN 0-9583834-1-3.

Annegarn, H. 2013. 'Potholes on the road to a Green Economy and what needs to be done to fill them' in L. Mujakachi, H. du Plessis and H. Annegarn (eds.), *Implementing Strategies Towards a Low-carbon Economy*. Proceedings of the Government Cluster Policy Workshop convened by the Human Sciences Research Council (HSRC), 4 March, Pretoria.

Annegarn, H., D. M. Guy and L. Mujakachi. 2010. 'Sustainable Human Settlement Development: The Witsand Housing Development in Atlantis, Western Cape Province'. Presentation made to the SALGA (Gauteng) Human Settlements Sub-Committee, 30 July 2010. SeTAR Centre, University of Johannesburg.

Annegarn, H., D. M. Guy and L. Mujakachi. 2011. 'Green Housing: Integrated Energy Efficiency, Empowerment and Cost Optimisation in Housing Delivery'. Presentation made at the launch of the West Rand District Municipality Green IQ. Muldersdrift, 16 May 2011. SeTAR Centre, University of Johannesburg.

Atkinson, D. 2002. *A Passion to Govern: Third Generation Issues Facing Local Government in South Africa*. Human Sciences Research Council and Centre for Development and Enterprise, June 2002.

Bega, S. 2015. 'Solar water heating on ice', *Saturday Star*, 24 January 2015. Available at: www.iol.co.za/business/companies/solar-heating-sector-on-ice-1.1808583 [Accessed May 2015].

Chown, D. 2014. 'Renewable Energy: The universe's complexity' in *Financial Mail: The Green Report 2014*, 30 October 2014. Available at: www.financialmail.co.za [Accessed July 2015].

Department of Cooperative Governance and Traditional Affairs. 1998. *White Paper on Local Government*, 9 March 1998.

Department of Energy. 1998. *White Paper on the Energy Policy of the Republic of South Africa*. ISBN: 0-9584235-8-X. December 1998.

Department of Energy. 2003. *White Paper on the Renewable Energy Policy of the Republic of South Africa*. General Notice 513 of 2004. *Government Gazette No. 26169*, November 2003.

Department of Energy. 2005. *Energy Efficiency Strategy of the Republic of South Africa*, March 2005.

Department of Energy. 2007. *Free Basic Alternative Energy Policy (Households Energy Support Programme)*. General Notice 391 of 2007, *Government Gazette No. 29760*, 2 April 2007.

Department of Energy. 2011. 'Electricity Regulation Act No. 4 of 2006: Electricity Regulations on the Integrated Resource Plan 2010–2030', *Regulation Gazette No. 9531*, Vol. 551. *Government Gazette No 34263*. 6 May.

Department of Energy. 2012. *Non-Grid Household Electrification Policy Guidelines*, 19 March 2012.

Department of Energy. 2013. *Integrated Resource Plan for Electricity (IRP) 2010–2030. Update Report 2013*, 21 November.

Department of Energy. 2015. Media Release: 'Solar Water Heater Programme Update'. Pretoria, 4 February 2015. Available at: www.energy.gov.za/files/media/pr/2015/MediaRelease-Solar-Water-Heater-Programme-Update-4February2015.pdf [Accessed May 2015].

Department of Environmental Affairs. 2008. *People-Planet-Prosperity: A National Framework for Sustainable Development in South Africa*, July 2008.

Dugard, J. 2009. 'Power to the people? A rights-based analysis of South Africa's electricity services' in D. A. McDonald (ed.) *Electric Capitalism – Recolonising Africa on the Power Grid*. Cape Town: HSRC Press. ISBN 978-0-7969-2237-3, pp. 264–287.

Economic Development Department. 2014. *Annual Report 2013/2014 of the Economic Development Department – Vote 28*. July 2014. Republic of South Africa.

Eltrop, L. 2011. 'The EnerKey Research Programme' in L. Mujakachi and H. Annegarn (eds.) *Proceedings of the Energy and Climate Protection Conference – A Reader*. Unpublished proceedings of conference held at Turbine Hall, Johannesburg, 5–6 October 2011. SeTAR Centre, University of Johannesburg.

Foxon, T. J. 2002. 'Technological and institutional 'lock-in' as a barrier to sustainable innovation'. ICCEPT Working Paper, November 2002. Available at: http://www.iccept.ic.ac.uk/public.html [Accessed April 2014].

Guy, D. 2010. *Witsand integrated Energy, Environment, Empowerment and Cost Optimization (iEEECO™) Benchmark Human Settlement*. Cape Town: PEER Africa.

Ledger, J. 2011. 'Sustainable energy supply – alternative energy technologies' in L. Mujakachi and H. Annegarn (eds.) *Proceedings of the Energy and Climate Protection Conference – A Reader*. Unpublished proceedings of conference held at Turbine Hall, Johannesburg, 5–6 October 2011. SeTAR Centre, University of Johannesburg.

Lilliestam, J., A. Battaglini, C. Finlay, D. Fürstenwerth, A. Patt, G. Schellekens and P. Schmidt. 2012. 'An alternative to a global climate deal may be unfolding before our eyes' in *Climate and Development*, Vol. 4, No. 1, January 2012, pp. 1–4.

Lund, T. 2015. 'Renewable Energy: Some light on horizon', *Financial Mail*, 23 April 2015. Available at: www.financialmail.co.za/2015/04/23/renewable-energy-some-light-on-horizon [Accessed May 2015].

Mapako, M. and G. Prasad. 2005. 'The Free Basic Electricity (FBE) policy and rural-grid connected households, solar home system (SHS) users and unelectrified households' in *Proceedings of Domestic Use of Energy Conference 2005*. Cape Peninsula University of Technology (CPUT).

Marquard, A. 2006. 'The Origins and Development of South African Energy Policy', Ph.D. thesis, University of Cape Town. January 2006.

Mohlakoana, N. 2014. 'Implementing the South African Free Basic Alternative Energy Policy – A Dynamic Actor Interaction', Ph.D. diss., University of Twente, Netherlands. December 2014.

Paraffin Safety Association Southern Africa (PASASA). 2012. 'The Need for a Household Energy Strategy and Policy – Submission to the National Planning Commission in response to the National Development Plan'. Report from the stakeholder workshop held 3 May 2012. Johannesburg.

Prasad, G. (n.d.): *South African Electrification Programme.* Energy Research Centre, University of Cape Town.

Presence, C. 2014. 'SA's "sad story of housing" – Lindiwe Sisulu', *Mail & Guardian*, 15 July 2014. Available at: http://mg.co.za/article/2014-07-15-sas-sad-story-of-housing [Accessed January 2015].

Presidential Infrastructure Coordinating Commission. 2012. *Provincial and Local Government Conference: A summary of the Infrastructure Plan*, 13 April 2012.

Republic of South Africa. 2011. 'Green Economy Summit Report'. 18–20 May 2010, Johannesburg. Available at: www.sagreeneconomy.co.za [Accessed May 2015].

Ruiters, G. 2009. 'Free basic electricity in South Africa: A strategy for helping or containing the poor' in D. A. McDonald (ed.) *Electric Capitalism – Recolonising Africa on the Power Grid.* ISBN 978-0-7969-2237-3, pp. 248–263. Cape Town: HSRC Press.

South African Bureau of Standards (SABS). 2011. *South African National Standard: Energy Efficiency in Buildings.* SANS 204: 2011 Edition 1, August. ISBN 978-0-626-26403-1. Pretoria.

Van Ryneveld, P., S. Parnell and D. Muller. 2003. *Indigent Policy: A Pro-Poor Income Strategy and Implementation Plan.* Prepared for the City of Cape Town, May 2003.

Verwey, B. 2014. 'Energy Efficacy and Efforts to Promote Green Living' in proceedings of Energy Efficient Housing Workshop held 29–30 September 2014, Cape Town. Workshop convened by the Renewable Energy and Energy Efficiency Partnership (REEEP), GreenCape and South African National Energy Development Institute (SANEDI).

SECTION III

Regional Integration: Exploring Optimal Energy Strategies

CHAPTER 10

Off-grid Renewable Electrification as a Viable and Complementary Power Planning Paradigm in Southern Africa: A Quantitative Assessment

Ogundiran Soumonni

ABSTRACT

The aim of this chapter is to present a quantitative assessment of some of the technological, economic and environmental implications of extending electricity to households in southern Africa through off-grid electrification as an alternative and complementary approach to the dominant centralised generation strategies. The first part of the chapter presents an overview of the literature on the environmental impact of various energy technologies over their entire lifecycles in order to provide a sense of scale with respect to the ecological advantage of renewable energy technologies (RETs). Based on meteorological data, the chapter then provides an estimate of both the theoretical and practical potential for distributed renewable electricity from solar and wind energy. Furthermore, using a 'component sizing' procedure, the chapter demonstrates how solar photovoltaics, and wind turbines (and their associated components such as batteries, charge controllers and inverters) can meet the basic electricity requirements of individual households in non-electrified areas. Lastly, the chapter presents the results of a set of simulations using the Hybrid Optimization Model for Electric Renewables (HOMER) software package to compare the costs of renewable electricity in a setting with little or no grid access, when compared to the more widely available diesel generator sets. The results strongly suggest that a decentralised, bottom-up and gradual approach to electrification is a viable and complementary policy paradigm for energy planning in southern Africa. If adopted in a manner that encourages ongoing learning, the approach should ultimately demonstrate more favourable implications for increased energy access, sustainability and technological competence building in the region and in other similar contexts.

INTRODUCTION

The objective of this chapter is to present a quantitative analysis of some of the key environmental and economic implications of extending distributed renewable electrification to residents of the southern African region not currently connected to the existing grid.

The southern African region, as used in this chapter, refers to the 15 countries that are part of the Southern African Development Community (SADC), which includes the 12 member countries of the Southern African Power Pool (SAPP) and three island nations that are not currently part of the regional power pool. Although the accuracy of international energy data varies by source, according to the SAPP, the average electrification rate of its member states is about 36 per cent (SAPP, 2014), with the International Energy Agency (IEA) reporting rates ranging from about nine per cent in Malawi and the DRC to 85 per cent in South Africa, and 97 per cent and 100 per cent in the island countries of Seychelles and Mauritius, respectively (IEA, 2014). Among other objectives, some of the key goals of the SAPP are to implement a competitive electricity market in the region, promote regional expertise through training and research, and increase access to electric power in rural communities (SAPP, 2015).

While the goals of the SAPP highlighted above are commendable, there is no explicit linkage between its planned energy generation and transmission projects and increasing energy access. It should also be noted that the eight generation projects highlighted in its 2014 annual report that received approval of their Environmental and Social Impact Assessments (ESIA) are based on large coal, natural gas and hydroelectric plants (SAPP, 2014). Since such large projects necessarily incur immense costs, which are typically funded by international financing institutions (IFIs), a prior study done on a sister institution, the West African Power Pool (WAPP), by Pineau (2008), suggested that the return on such investments is typically seen as being unlikely to come from rural and peri-urban residents whose generally low income and minimal consumption of electricity are unlikely to justify such expenditures. Similarly, a more recent comparative assessment of the Grand Inga Dam project in the Democratic Republic of Congo (DRC), which is expected to generate about 39,000MW at a cost of about $US80 billion, suggests that, with respect to affordability, the national debt burden, social and environmental sustainability and innovation, the outcomes of the project are likely to be negative for the country, though they might be beneficial for large players in the extractive industry (Green, *et al.* 2015). Lastly, a compelling study by Levin and Thomas (2011) employs spatial modelling based on a network algorithm to determine the conditions under which centralised or decentralised electrification can be provided at least cost across the globe, and demonstrates that, for most of the world's population, especially in Africa, large regions can be served by decentralised electrification at low cost. Levin and Thomas (2011) emphasise that the advantages of distributed electricity include an installation time of days instead of years or even decades for grid-based electricity, accessibility and reliability, which could make it attractive even in countries that have high population densities such as Bangladesh or Nigeria.

The three examples cited above are reflective of a wider literature that argues that large, centralised projects in many less industrialised countries (this is typical for sub-Saharan Africa), tend to guarantee the reliability of supply for large industrial, commercial and urban consumers, thereby leaving the problem of sustainable access generally unresolved. Consequently, the critical premises of this chapter situate it within that body of work. However, there are a few exceptions to this pattern of centrally-planned power generation only serving a minority of residents across sub-Saharan Africa, the most notable of which was South Africa's post-1994 electrification programme which, although primarily reliant on coal-based power, nevertheless increased the country's electrification rate from about 30 per cent in 1990 to over 80 per cent in 2007 (Bekker, *et al.* 2008). While the small island nations of Seychelles and Mauritius have also achieved high electrification rates, the general situation in other SADC countries (except Botswana and South Africa) is that they all have electricity access rates of less than 50 per cent. As shown in Table 1 below, 10 of the 15 SADC states have electrification rates of 30 per cent or less.

Table 1: Electricity Access and Consumption (KWh) (IEA, 2014)

Country	Electricity Access Rate (%)	Electricity Consumption Per Capita (kWh)
Angola	30	248
Botswana	66	1,603
D.R.Congo	9	105
Lesotho	28	-
Madagascar	15	-
Malawi	9	-
Mauritius	100	-
Mozambique	39	447
Namibia	30	1,549
Seychelles	97	-
South Africa	85	4,606
Swaziland	27	-
Tanzania	24	92
Zambia	26	559
Zimbabwe	40	757

Following the brief description of the electricity access situation above, the focus of the chapter is to demonstrate the feasibility of an alternative planning approach for increasing the rate of sustainable electricity in southern Africa, namely through the distributed generation of renewable energy and, more specifically, off-grid electrification as its basic building block.

Distributed generation (DG) or decentralised generation typically refers to small-scale technologies that generate power on the order of 10kW up to 50MW in the vicinity of a consumer, and which may include both renewable systems such as solar panels, biomass generators or wind turbines, and non-renewable systems such diesel generators or combined heat and power systems (CHP) (Sovacool, 2008). DG can also be defined as a class of small-scale electrical power generation technologies that provide electric power at a load site, or adjacent to one, and that can either be directly connected to a consumer's facility, to a distribution system, or both (Borbely and Kreider, 2001). The chapter therefore sets out to illustrate how this planning approach might work in southern Africa from the point of view of increased access, sustainability of supply, costs, and environmental benefits. Although the planning concept may apply to various renewable sources, the chapter focuses on solar and wind energy, either in combination with diesel engines, or as stand-alone sources of power.

The chapter begins the assessment with a quantitative representation of the current demand and supply of power in southern Africa as provided by the authorities of the Southern African Power Pool (SAPP). Secondly, it provides an estimate of both the theoretical potential and a more operational or practical potential for distributed renewable solar and wind energy, based on the available meteorological data and explicit assumptions applied to a spreadsheet model. Using a 'component sizing' procedure that allows consumers to tailor their energy generation capacity to their needs and financial means, the chapter further demonstrates how solar photovoltaics, wind turbines, biomass gasification generators, and their associated components such as batteries, charge controllers, inverters, or gasifiers, could meet the basic electricity requirements of individual households in non-electrified areas. The basis for modelling these minimum requirements is drawn from the electrification experience of South Africa, which implemented a Free Basic Electricity (FBE) Policy that required that every indigent or poor household be provided with 50kWh a month free of charge in order to stimulate residential electricity consumption (Bekker, *et al.* 2008). While some provinces such as KwaZulu-Natal have raised this figure to 100kWh, certain environmental groups have lobbied for the FBE to be raised to 200kWh, either through grid-based electricity or through distributed resources such as solar home systems (SHSs) for remote areas (Earthlife Africa, 2010). In the case of indigent households, Free Basic Alternative Energy (FBAE) subsidies are in existence, which include SHSs as well as non-electricity sources such as paraffin, bio-ethanol gel, liquefied petroleum gas and coal (RSA, 2007). Finally, the chapter presents an analysis of the comparative costs of renewable electricity in a setting with no grid access, when compared both to the more widely available diesel generator sets and grid-based electricity, through simulations with the Hybrid Optimization Model for Electric Renewables (HOMER) software package.

Figure 1: Installed Capacity by Technology (SAPP, 2014)[82]

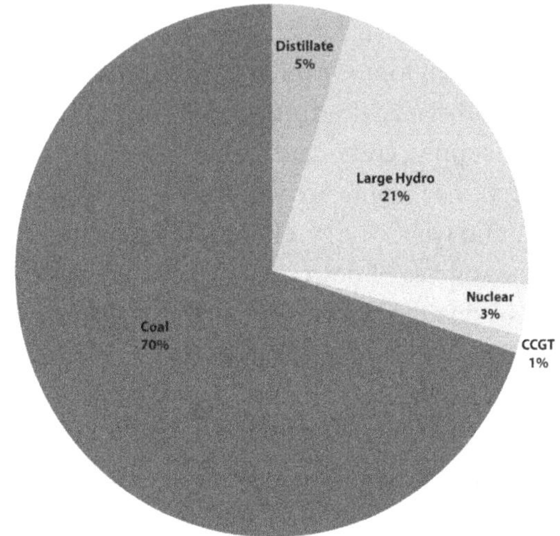

STATUS OF POWER DEMAND AND SUPPLY IN SADC

The most recent data for electricity generation in southern Africa by the SAPP was made publicly available in 2014, but was released in terms of installed generation capacity in MW, which does not tell us how much was actually generated as a result of maintenance or related issues. Thus, the data on the actual electricity generated, though dating to 2012, was acquired from the International Energy Agency (IEA), which tracks this data globally, and was deemed to be more appropriate for the analysis. The figures are primarily from centralised generation (both renewable and non-renewable sources), since electricity from distributed generation in SADC is considered to be negligible and the corresponding energy data is not yet tracked in most of its member states. Figure 1 shows the relative share of various energy sources across the region by installed capacity and reveals that coal is dominant, followed by large hydroelectric power, distillate (oil and diesel),

82. CCGT – Combined Cycle Gas Technology; Distillate – Distillate Fuel Oil.

nuclear power, and finally, energy from combined cycle gas turbine (CCGT) technology.

Table 2: Net Electricity Generated in SADC (GWh) (IEA, 2012)

Country	Thermal	Nuclear	Hydro	Biofuels	Renewables (Solar + wind)	Total (GWh)
Angola	1,633	-	3,980	-	-	5,613
Botswana	250	-	-	-	-	250
D. R. Congo (DRC)	35	-	7,931	-	-	7,966
*Lesotho	-	-	-	-	-	486
*Madagascar	-	-	-	-	-	2,025
*Malawi	-	-	-	-	-	2,180
Mauritius	2,218	-	74	500	-	2,792
Mozambique	21	-	15,145	-	-	15,166
Namibia	36	-	1,607	-	-	1,643
*Seychelles	-	-	-	-	-	320
South Africa	239,538	13,075	4,860	293	153	257,919
*Swaziland	-	-	-	-	-	425
Tanzania	4,106	-	1,657	19	-	5,795
Zambia	36	-	11,814	-	-	11,850
Zimbabwe	3,638	-	5,387	63	-	9,088
Total (GWh)						323,518

*Note: No specific or general data for Lesotho, Madagascar, Malawi, Seychelles and Swaziland was provided by the International Energy Agency (IEA) and therefore, the numbers for 'Total Electricity Generated' were based on the electricity generation data reported in 2012 by the US Energy Information Administration (EIA).

Table 2 shows the net energy generated per country and by source, and highlights the fact that only South Africa, as of 2012, generated utility-scale solar and/or wind power, while Mauritius, Zimbabwe, Tanzania and South Africa all generated minimal to modest amounts of electricity from biomass.

It should be noted that in this chapter, only solar, wind, biomass power from waste, and pico-, micro- and mini-hydroelectric technologies are considered to be renewable. Large hydroelectric dams are not considered renewable energy technologies (RETs) because they tend to radically alter the flow of the river that has been dammed, flooding large areas of land that include farm land, buildings and other property, or wildlife habitats, as well as routinely displacing people who live downstream from the dam. On the other hand, small dams, also known as run-of-the-river schemes, only divert a small part of the river to produce electricity due to the gravity upstream, but the water later flows back into the river, thereby significantly reducing land use impacts. Furthermore, the water requirements of large dams, which create artificial lakes with stagnant water, are highly competitive with other critical needs such as drinking, fishing, sanitation and recreation, among others. This typically implies the diminished use of the river for these purposes, unless power production itself is reduced and balanced with other needs. Finally, a review of hydrological studies between 1950 and 1989 has shown a disruption in rainfall patterns since 1970 in the case of West Africa (Gauthier, *et al.* 1998), which currently manifests itself as a reduction in water levels in rivers and dams in the region that adversely affect electric power generation. Thus, the combination of factors highlighted above seriously undermines the notion of large hydroelectric dams being based on an inexhaustible or renewable resource.

LIFE CYCLE ASSESSMENT OF TECHNOLOGY OPTIONS: A REVIEW

The purpose of this section is to provide a brief technology assessment of the existing options for electrification in southern Africa based on a review of the literature. It is necessary to make this case in this chapter because a number of criticisms are widely levelled against renewable energy technologies in policy discourse that are without adequate evidence of specification. For instance, it is often stated that the challenge with renewable energy technologies is their supposedly high cost, without clear reference to either capital costs or lifecycle costs, or price distortions such as the absence of carbon taxes on fossil fuels or compensatory subsidies for renewables. Another source of contention relates to diverging interpretations about which technologies are more environmentally beneficial than others, depending on the point at which they are assessed (either at the end-use or some other unspecified set of points). For instance, some commentators argue that the manufacturing of solar PV panels is so energy-intensive that the gains in terms of pollution are minimal over conventional fossil-based generation, while others argue for large-scale biofuels investments on account of their lower emissions at the point of use, without necessarily accounting for the loss of trees and plant cover from the land clearing that is needed to grow energy crops. Thus, in either case, there is a lot of doubt in the public mind as to whether renewable energy solutions are even beneficial in the long run with respect to their energy intensity during the manufacturing phase, their reliability and other factors. While the systematic cost comparisons are presented in the fifth section, the technology assessment provided here offers a review of quantitative comparisons of these technologies, which have been done across their entire life cycles.

A life cycle can be defined as the total life span that a product goes through from the extraction of the raw material used to produce it, its manufacture, its use and, ultimately, to its disposal as waste into the environment (Nieuwlaar, 2004). A Life Cycle Assessment (LCA) then refers to the evaluation and analysis of the life cycles of various products. While an LCA is typically limited to environmental issues, the term can sometimes include social and economic considerations as well (Nieuwlaar, 2004). However, the LCA is used in this chapter only in reference to environmental issues, and is distinguished from the more widely known Environmental Impact Assessment (EIA) because of the emphasis of the LCA on the products themselves. The results obtained from a levelised cost of electricity (LCOE) analysis, which is a standard analytical technique for comparing the life cycle costs of different systems (Byrne, *et al.* 2007), are subsequently presented in the HOMER modelling exercise in the fourth section. Furthermore, the net present costs (NPC) of the optimised technology options presented in that section are the same as the life cycle costs according to developers of the HOMER software package that was used in the modelling exercise.

Although there are methodological debates around life cycle assessments, they go beyond the scope of this chapter, and the goal of this section, then, is simply to use examples from the peer-reviewed literature to demonstrate the types of materials and pollutants, as well as their general orders of magnitude that should be accounted for in selecting a particular energy system. The current section sheds more light on the LCA characteristics of the main energy systems used in much of sub-Saharan Africa, more generally, for both centralised and distributed generation (DG). The centralised generation technologies are primarily based on hydropower, oil, natural gas and, to a lesser extent, biomass power, while the DG technologies include diesel generators, solar photovoltaic panels, wind turbines and biomass gasification systems. With respect to LCA, the difference between fossil fuel-based energy systems and non-fossil-based sources, including nuclear energy, is that in the former, the main environmental impacts come from combustion and their associated waste and flue gases, while in the latter, the impacts are based on the environmental profile of the materials used in their manufacture (Nieuwlaar, 2004).

The life cycle assessment described below of a 3kW community hydroelectric power system that was juxtaposed with assessments of a diesel generator and electricity from the grid in Thailand can be compared favourably to the sub-Saharan African context in general, and large areas of southern Africa as well. In that assessment, Pascale, *et al.* (2011) demonstrated that when placed in a rural setting, the 3kW run-of-the-river generator produced better environmental and financial outcomes than both a diesel generator and grid-based power generated from various sources, including: natural gas (72%), coal (15%) and large dams (7%). Their result was in contrast to most other LCA studies that had previously shown that smaller hydroelectric generators have a higher environmental impact per kWh than larger grid-based systems, which shows the site-specific dependence of some LCAs due to the use of different assumptions and considerations. Some of the inputs to the hydropower system located in rural Thailand included the associated components of the 3kW pump itself, namely the motor, capacitors, induction generator controller and other inputs such as PVC pipe parts or galvanised steel parts, all measured in kilograms. In addition, the hydropower system outputs included forest use change (m^2), potential riverbed use change (m^2), and the lifetime energy available at the connection to the buildings (kWh), among others (Pascale, *et al.* 2011).

Furthermore, Pascale, *et al.* (2011) measured seven different categories and indicators in their life cycle impact assessment (LCIA), which is a phase of the LCA that seeks to evaluate the magnitude and significance of potential environmental impacts on a product. These included the global warming potential (GWP) over 100 years, the acidification potential (AP), the eutrophication potential (EP), the ozone layer depletion potential (ODP) at steady state, the petrochemical ozone creation potential (POCP), the abiotic depletion potential (ADP) and the primary energy demand (PED) from both renewable and non-renewable resources. The unit of measurement of the last indicator is in kWh, but the first six indicators are measured in g/kWh because 1kWh is the functional unit or reference unit for the quantified product performance that is typically used in most LCA studies on electricity generation (Pascale, *et al.* 2011). Six of these categories (except PED) are consistent with the nine recommendations of the US Environmental Protection Agency (EPA) for assessing energy systems with the three additional categories being eco-toxicity, human toxicity, and land use (Nieuwlaar, 2004).

Table 3 below shows the total environmental impacts of the community hydroelectric generator across its entire life cycle, as well as the corresponding impacts from a diesel generator and grid-based power, where the indicators for the latter two technologies are represented as multiples of those of the hydro-powered generator, which is used as the baseline. It covers six steps within the product system including the intake or dam, the penstock or the pipe that takes the water to the turbine, the turbine itself, the transmission line, the control house and, ultimately, the distribution component, with transmission having the highest impact across most of the categories.

Table 3: Total LCIA Results for 3 kW Community Hydroelectric Generator Compared to Diesel and the Grid (Pascale, *et al.* 2011).

Category or Indicator	Hydroelectric generator (g/kWh) – baseline	7kV A diesel generator	Grid-based power connection
ADP (g Sb-e)	0.264	x(45)	x(23)
AP (g SO_2-e)	0.372	x(9)	x(29)
EP (g PO_2-e)	0.030	x(13)	x(14)
GWP (g CO_2-e)	52.7	x(27)	x(18.5)
ODP (g R11-e)	$3.13 * 10^{-6}$	x(1.5)	x(4)
POCP (g ethene-e)	0.030	x(9)	x(22)
PED (kWh)	0.150	x(46)	x(24)

Many studies that perform LCAs of renewable energy technologies tend to focus almost exclusively on greenhouse gas (GHG) emissions such as CO_2, CH_4, and N_2O as the measure of environmental impact because these gases have the greatest impact on climate change, and because alternative energy industries consider this to be the most relevant metric that can prove that their technologies have a lower impact on global warming (Fleck and Huot, 2009). In this regard, the LCA approach is advantageous because it helps to determine whether or not this assertion is verifiable and serves to clarify the extent to which RETs may outperform conventional technologies from an environmental point of view. It does this by including both the emissions from upstream processes such as energy resource extraction or the technology production, which are usually neglected in other analytical methods, and the downstream emissions from the electricity generation stage. While the GHG emissions from fossil-fuelled technologies mostly come from downstream processes, and make up more than 75 per cent of the direct emissions from the plant, emissions from renewable energy technologies are dominated by the upstream processes that usually account for more than 90 per cent of their cumulative life cycle emissions (Weisser, 2007).

Table 4: GHG Emissions from Electricity for Selected Energy Technologies. Adapted from (Cherubini, *et al.* 2009).

Energy technology system	GHG emissions (g CO_2-eq./MJ)[83]
Biomass	15–30
Biogas	15–65
Wind	1–10
Hydro	0.5–10
Solar PV	15–40
Coal	300–500
Oil	200–300
Natural Gas	100–200

A study on the greenhouse gas-based LCA of biofuel and bioenergy systems found that electricity generated from biomass residues typically does not increase pressures on the environment because those wastes are not specifically produced to be used as energy resources (Cherubini, *et al.* 2009). There are exceptions to this, however, such as first, if the extent of removal of residues from land depletes the soil nutrients or reduces the carbon stored in carbon pools like soil or dead wood, and secondly, if the creation of a market for residues leads to the production of the original commodity, such as timber. In terms of GHG emissions per unit of output of energy, the

83. The equivalent carbon dioxide (CO_2-eq.) unit represents the amount of a given greenhouse gas (GHG) or mixture of GHGs that would cause the same level of radiative forcing, that is, that would have the same global warming potential (GWP) as an equivalent amount of CO_2.

emissions that result from either firing biomass directly or from the biogasification of biomass are shown in comparative perspective with other selected renewable energy and fossil fuel technologies in Table 4.

The data in Table 4 shows that the GHG emissions from biomass and biogas-generated electricity are larger than those of wind and hydropower, and are comparable to those of solar PV. However, the emissions from fossil-fuelled generators are higher than those from renewables by an order of magnitude. The ranges in data for a given technology can be attributed to datasets from different sources because different LCAs rely on life cycle inventories with different input values, and can be explained by the different allocation rules applied to bioenergy systems because many of them produce different material products and co-products relating to either electricity or heat, or both (Cherubini, *et al.* 2009).

Finally, a direct comparative study between the life cycles of a wind turbine and a solar PV module by Zhong, *et al.* (2011) confirmed that the former has lower GHG emissions than the latter, but it also demonstrated that wind turbines have a lower environmental impact across all the assessment categories. The authors found that the larger amount of fossil fuels used in the energy-intensive manufacturing process of solar PVs had the most significant impact on the difference in GHG emissions between the two technologies. However, they also argued that even though it would be beneficial for energy from solar PVs to be used to help power the assembly phase, a bigger challenge would remain in the disposal phase because only small quantities of the recycled waste can currently be reaped from solar PVs as compared to wind turbines. In the case of a wind turbine system, Fleck and Huot (2009) found that the most emission-intensive process was in the production of the battery banks, followed by the tower production, and thirdly, battery bank delivery, with the turbine production process itself and all other processes accounting for only eight per cent of the system's GHG emissions.

In conclusion, this section has undertaken to clarify the appearance of equally viable contradictory views about the ecological viability of renewable energy technologies over conventional energy sources. It has shown that the preponderance of evidence based on life cycle assessments is in favour of RETs over fossil-based distributed sources, as well as grid-based electricity from large hydroelectric dams and fossil-fuelled generators.

ESTIMATION OF RENEWABLE RESOURCES

Following the case made in the preceding section, this section then seeks to address another popularly expressed scepticism about the practical availability of renewable energy sources. It thereby illustrates the abundance of these resources, specifically with respect to meeting the electricity requirements of the region. The renewable resources chosen in this section are based on a cursory view of meteorological maps that suggest their availability, and the assessment of these resources is deemed sufficient to support the argument that is being advanced in this chapter.

Solar Energy

Solar energy technologies can exist as either solar thermal systems or solar photovoltaics (PV). The first set of technologies focus sunlight on a fluid to produce steam that can then be used in a turbine to generate electricity, while the latter convert sunlight directly into electricity. I focus here on the energy potential from solar home systems (SHSs), as these are more suitable for residential use. SHSs typically include the PV modules, batteries, a charge controller and an inverter. However, it should be noted that while an inverter is primarily needed only if one intends to use appliances that operate with Alternating Current (AC), it is also a useful device to have in order to provide the flexibility to connect with an existing or future centralised grid.

The approach employed to obtain the meteorological data for assessing the solar potential was obtained in the following manner. In order to find geographical coordinates for the countries in the region, the Google Earth software was used to select two diagonally located cities or towns on a given country's map so as to increase the chances of capturing any significant climatic contrasts in that particular country and, consequently, in the region overall. The average insolation incident on a horizontal surface was then acquired by entering each coordinate into a database on surface meteorology and solar energy that is maintained by the Atmospheric Science Data Center of the US government's National Aeronautics and Space Administration (NASA) database. The software associated with the database then derives the average solar radiation resolved over the area between the two locations for the user. The latitude, longitude, elevation and corresponding resolved average solar insolation data (collected over a 22-year period from July 1983 to June 2005) of the selected locations for analysis are displayed in Table 5 below.

Estimation of Electricity Potential from Solar Photovoltaics

The goal of the first part of the calculations in this sub-section is to illustrate that solar energy can meet the electricity requirements of the region, as well as show how various sizes of solar PV modules could meet the minimum electricity requirements of a non-electrified household. Based on the characteristics of current solar technologies, the solar resource in southern Africa is shown to be sufficient to meet all its electricity needs many times over. This kind of expansive assessment is frequently done to demonstrate the technical feasibility of using a given resource. For example, Sobin (2007) relies on several seminal studies that show that the amount of solar energy that the US receives is comparable to 500 times its energy demands, and that about 0.4 per cent of its area (or 10 million acres) could provide all the electricity consumed in the country using current PV technologies, in order to refute the myth that renewable energy can never meet the electricity demand in the US. For the case of southern Africa, the number of PV modules is calculated that would be needed to meet minimum household electricity requirements of 50kWh, 100kWh and 200kWh in each country, given its average solar radiation resolved over a large area of the country. The calculation used to estimate the amount of electricity from solar energy that can be generated from the land area available in each southern African country is shown below for the case of Botswana. The solar power that can be produced from a particular land area is given by:

$$Power = Area * S_{avg} * \eta \qquad \text{(Hafemeister, 2007)}$$

where :

$$S_{avg} = Average\ solar\ flux \left(\frac{kW}{m^2}\right) = Avg.\ solar\ radiation \left(\frac{kWh}{m^2\,day}\right) * \left(\frac{1\ day}{24h}\right)$$

η=*Efficiency of Solar PV cells*

Area = Land area in a given country (m^2)

The solar energy produced in one year can then be calculated as:

$$Energy\ Generated = Solar\ Power * 8{,}760\ hours = (Area * S_{avg} * \eta) * 8{,}760\ hours$$

Thus, in one year in Botswana:

$$Energy\ Generated = [5.82*10^{11} m^2 * (6.362 \left(\frac{kWh}{m^2\,day}\right) * \left(\frac{1\ day}{24hrs}\right) * 0.15] * 8{,}760\ hrs$$

$$= 1.29 * 10^{15}\,kWh = 3.16 * 10^{7}\,GWh$$

This amount is about 5.2 million times the total amount generated (250GWh) in Botswana in 2012 (see Table 3). More specifically, if that amount of 250GWh were to be generated from solar energy it would represent only 0.00002 per cent (or 0.12km^2) of the total land mass of the country. Depending on the country's size and its electricity demand, there is a wide variation in this hypothetical measure, wherein South Africa would need to dedicate the equivalent of 0.01 per cent (or 122.1km^2) of its land to solar energy, whereas Seychelles would need 0.14 per cent (or 0.64 km^2).

The calculations that follow below are then used to estimate a more practical policy, that is, the amount of solar energy that can be generated to meet the minimum monthly requirements of the Free Basic Electricity policy in South Africa, which is applied to the entire region and hereby referred to as 'Basic Electricity' (BE). As explained in the introduction, it is important to recall that this minimum requirement is not adequate for a typical household, but as in the centralised case, it does ensure that households have access to electricity, after which they can increase their consumption to the extent that they can afford to. Indigent households that are connected in this way will have to use non-electric sources for heating and cooking, such as paraffin or LPG – the latter already widely used for cooking and heating even in wealthier households.

Four sizes of polycrystalline PV modules are considered ($50W_P$, $100W_P$, $150W_P$, $250W_P$), which are available on the market and are suitable for use in Solar Home Systems (SHSs). The number of panels of each type that would be needed to meet the three possible minimum levels of electrification is then calculated. As Swaziland's annual solar insolation is very close to the regional average for SADC countries, that country's data for the purpose of demonstration is used. The results are shown in Table 5.

The method used for the estimation of the number of PV panels that would be needed to meet monthly basic electricity (BE) requirements of 50kWh, 100kWh and 200kWh, is shown as follows:

The capacity factor (CF) of a solar PV panel, which is determined from the power rating on the panel at a peak solar flux (S_{peak}) of 1 kW/m^2, is given by:

$CF_{PV} = S_{avg}/S_{peak}$ (Hafemeister, 2007)

The solar energy generated by one panel is calculated as:

Energy Generated = *Average Solar Power* $*$ *30 days* $*$ *24 hours*
= Average Solar Power $*$ *720 hours*
= *(Power rating* $*$ *CF* $*$ *Panel loss factor)* $*$ *720 hours*

In this calculation, the panel loss factor includes a multiplier to account for operating temperature (0.9), a multiplier to account for losses through the cables (0.98), the efficiency factor of an inverter (0.85), and an installation factor, which accounts for the incline, orientation and coatings on a PV panel (1.04) (Coley, 2008).

Thus, for an average insolation of 5.29 kWh/m²day in Swaziland,

$$CF_{PV} = \frac{S_{avg}}{S_{peak}} = \frac{\left(5.29 \frac{kWh}{m^2\,day} * \frac{1\ day}{24hrs}\right)}{1\ kWh/m^2} = 0.22$$

The energy generated from one $50W_p$ (or $0.05kW_p$) in one month is calculated as:

Energy Generated = (Power rating ∗ CF ∗ Panel loss factor) ∗ 720 hours =

$$(0.05kW * 0.22 * 0.98 * 0.9 * 0.85 * 1.04 * 720\ hours = 6.19\ kWh$$

Finally the number of $50W_p$ panels required to meet a BE of 50kWh is:

$$No.\ of\ panels = \frac{Total\ energy\ generated}{Energy\ generated\ by\ one\ panel} = \frac{50kWh}{6.19kWh} = 8.08 = \sim 8\ panels$$

This demonstration can then easily be extended to the other power ratings and BE requirements in the case of Swaziland as shown in Table 5 below.

Table 5: Number of PV Modules Needed to Meet Monthly Basic Electricity Requirements

	50kWh/mo.	100kWh/mo.	200kWh/mo.
$50W_p$ panel	8	16	32
$100W_p$ panel	4	8	16
$150W_p$ panel	3	5	10
$250W_p$ panel	2	3	6

In steps 1 to 11 in Table 6 below, the number of batteries that would be needed to store the energy required to meet each of the monthly electricity requirements is calculated. The total number of batteries is shown in step 11, which indicates that two, four and six batteries would be needed to meet monthly requirements of 50kWh, 100kWh and 200kWh respectively.

Note: The specifications for the selected battery type above were obtained from the websites of the vendors, as well as the heuristic for calculating the number of required batteries. The calculation also assumes six hours of battery power needed per day, that is, in the evenings/early morning.

Table 6: Number of Batteries Needed to Meet Basic Electricity Requirements

	Measure*	50kWh/mo.	100kWh/mo.	200kWh/mo.
1	Daily Amp-hr Requirement ($I*hr=P*hr/V_{system}$); V_{system}=12 V	34.7	69.4	139
2	# of days of autonomy (# of consecutive days of cloudy weather)	1	1	1
3	# of Amp-hrs battery needs to store (1 * 2)	34.7	69.4	139
4	Depth of discharge	0.5	0.5	0.5
5	Effective Amp-hr requirement (3 ÷ 4)	69.4	139	278
6	Ambient temperature multiplier (at 80 F)	1	1	1
7	Total battery capacity needed (5 * 6)	69.4	139	278
8	Amp-hr rating of battery (12V SONX RT12260D Deep Cycle Battery)	26	26	26
9	# of batteries wired in parallel (7 ÷ 8)	2.67 (~3)	5.34 (~6)	10.68 (~11)
10	# of batteries wired in series (Nominal V ÷ Battery V)=12V ÷ 12 V	1	1	1
11	**Total # of batteries required (9 * 10)**	**3**	**6**	**11**

Wind Energy

In this section, the potential of wind energy to meet the electricity access requirements of the region is assessed. Wind turbines convert the kinetic energy of the flowing wind into electricity. The wind speed varies by height,

Table 7: Wind Energy: Geographical Coordinates and Average Wind Speeds (NASA, 2015)

Country	City	Latitude	Longitude	Elevation (m)	Resolved average wind speed at 10m (m/s)	Resolved average wind speed at 50m (m/s)
Angola	Luanda	13.26	8.83	59	--	--
	Cuando Cubango	18.8	16.33	1,221	4.43	5.60
Botswana	Gaborone	25.89	24.65	1,159.26	--	--
	Maun	23.42	19.98	942	4.51	5.70
D. R. Congo	Lubumbashi	27.5	11.67	1,303	--	--
	Kinshasa	15.32	4.33	351	4.47	5.66
Lesotho	Maseru	29.31	31.98	1,642	--	--
	Mokhothong	29.07	29.29	2,209	3.78	4.79
Madagascar	Antananarivo	47.53	18.93	1,352	--	--
	Itampolo	43.95	24.68	51	3.39	4.29
Malawi	Lilongwe	33.77	13.98	1,135	--	--
	Kaphiika	34.09	10.45	692	4.77	5.79
Mauritius	Port Louis	57.52	20.17	85	--	--
	Mahebourg	57.7	20.4	44	6.33	7.40
Mozambique	Maputo	32.56	26.01	22	--	--
	Tete	33.6	16.17	270	4.90	5.83
Namibia	Windhoek	17.11	22.6	1,723	--	--
	Okahandja	16.92	21.94	1,578	4.73	5.99
Seychelles	Victoria	55.45	4.62	72	--	--
	La Digue	55.83	4.36	206	9.67	11.31
South Africa	Cape Town	18.42	33.9	-1	--	--
	Thohoyandou	30.48	22.95	576	4.32	5.47
Swaziland	Mbabane	31.13	26.32	1,335	--	--
	Hluti	31.59	27.22	519	4.73	5.53
Tanzania	Dar es Salaam	39.28	6.78	23	--	--
	Mwanza	32.9	2.51	1,187	4.28	5.24
Zambia	Lusaka	28.28	15.43	1,185	--	--
	Kasama	31.2	10.23	1,365	4.10	5.19
Zimbabwe	Bulawayo	28.62	20.12	1,211	--	--
	Harare	31.03	17.85	1,351	4.29	5.32

and the wind speeds at 10m and at 50m for selected locations in southern Africa are shown in Table 7.

Estimation of Electricity Potential from Wind Turbines

According to Wood (2011), the primary factors that should be taken into consideration in order to estimate the electricity generation potential from wind at a particular location include the following: site assessment, which consists of mapping the wind resource and choosing ideal locations for the turbines, and optimal height of wind towers and installation loads during the raising and lowering of the turbines. The electrical power that can be generated by a wind turbine is given by the equation below:

$P_{wind} = 0.5\eta\rho v^3 A_{wind}$ (Hafemeister, 2007)

where:

η = efficiency of the windmill = ~25%

ρ = density of air = 1.293kg/m^3

v = average velocity of wind

A_{wind} = Swept area of turbine with rotor diameter (d) = $\pi(d/2)^2$

The energy generated by such a wind turbine then, is given by:

$Energy_{wind} = P_{wind} * hours\ of$ operation

From the equation for wind power above, we can observe that other than the wind speed, the rotor diameter is the most important determinant of the power that can be produced by a wind turbine. The rotor diameters that are typically used for wind turbines are 1.5m (micro) turbines used on yachts, for instance, 2.5m (mid-range) normally used for single-user remote or grid-connected households, and 5m (mini) used for mini-grids or remote communities (Wood, 2011).

Even though the estimation of wind potential is very model dependent, it can roughly be calculated by assuming the use of small wind turbines placed at an average height of 10m or they can be mounted higher at 50m and calculated over a given area. However, if wind speeds are too low, that is, below the *cut-in speed*, the turbine will not produce any power even though it may appear to be spinning. Wood (2011) discusses site assessment for wind turbines and states that as a rule of thumb, a wind speed of 5m/s is considered to be a good value at an average height. For a small 0.4KW generic wind turbine (rotor diameter = 4m), the *cut-in speed* is 3m/s. Given that all

SADC countries have average wind speeds that are at least 3.39 m/s or more at a hub height of 10m, for the selected sites the corresponding potential power generation is justifiable in this calculation. If we consider the country of Mozambique, for instance, with a land area of 799,380 km^2 or 4.03 $*$ 109 m^2, then the wind volume is:

$$V_{wind} = A_{wind} * height = 7.99 * 10^{11}m^2 * 10m = 7.99 * 10^{11}m^3$$

Thus, if we assume the use of windmills that operate at an efficiency of 25 per cent (note that the Betz limit or theoretical maximum is 59 per cent), the wind power that can be produced in Mozambique, with an average wind velocity of 4.9 m/s at a 10 m height is:

$$P_{wind} = 0.5\,\eta\rho v^3 V_{wind} = 0.5 * 0.25 * \frac{1.293kg}{m^3} * \left(\frac{4.9m^3}{s}\right) * 7.99\,10^{12}m^3 = 1.517 * 10^{14}W_e$$

$$= 151{,}723GW$$

The onshore wind energy potential from Mozambique in one year at 10m can then be calculated as:

$$Energy_{wind} = P_{wind} * hours\ of\ operation = 151{,}723GW * 8{,}760\ hours = 1.33 * 10^9GWh$$

At 50m, where the wind speed is 5.83m/s, the wind energy potential from Mozambique would be equal to 1.12 $* 10^{10}$ GWh, which is higher by an order of magnitude.

At 10m, the wind energy potential is about 87,636 times the met demand of 15,166GWh in Mozambique, while at 50m it is about 739,765 times the demand. As with the solar energy resource in the previous section, the analysis above shows that there is also a sufficient wind energy resource to meet the electricity needs of the region in principle. More specifically, the amount of land that is needed to produce Mozambique's current electricity generated is 0.001% (or 7.99km^2) at a height of 10m and 0.0001% (0.8km^2) at a height of 50m.

In order to demonstrate the consumption of this energy resource in practice, the three Basic Electricity (BE) scenarios of 50kWh/month, 100kWh/month and 200kWh/month as in the previous section are used, and how the small wind turbine rotor dimensions of 1.5m, 2.5m, and 5m could meet this demand is estimated. For a wind turbine with a rotor diameter of 2.5m, for example, the electricity produced in Mozambique with an average annual wind speed of 4.9m/s at 10m is:

$$Pwind = 0.5\eta\rho v^3 A_{wind} = 0.5\eta\rho v^3\pi(\frac{d}{2})^2 = 0.5 * .25 * \frac{1.293\ kg}{m^3} * (\frac{4.9}{s})^3 * \pi(\frac{2.5m}{2})^2 = 93.168W_e$$

The energy that can be generated in one month is then calculated as:

$$Energy_{wind} = P_{wind} * hours\ of\ operation = 93.168\ We * 30\ days * 24\ hours = 67{,}081Wh = 67.081kWh$$

Thus, the number of turbines needed to meet a BE of 50 kWh per month is:

$$No.\ of\ turbines = \frac{Total\ energy\ generated}{Energy\ generated\ by\ one\ turbine} = \frac{50kWh}{67.081\ kWh} = 0.54 = \sim 1\ turbine$$

This calculation is extended to the other rotor diameters and for all the BE requirements for southern Africa as a whole and the results are shown in Table 12. The number of batteries needed to meet the BE requirements is approximately the same as that shown in Table 8 below. Thus, two, four and six batteries are needed in combination with the turbines to produce BE requirements of 50kWh, 100kWh and 200kWh respectively.

Table 8: Number of Wind Turbines per Household Needed to Produce Minimum Electricity Requirements

	BE = 50kWh/mo.	BE = 100kWh/mo.	BE = 200kWh/mo.
1.5m turbine	2	3	6
2.5m turbine	1	1	2
5m turbine	1	1	1

The analyses above have revealed some of the technology and resource combinations by which solar and wind energy could increase energy access and meet minimum basic electricity requirements in southern Africa. The amount of energy that would need to be generated for each country to meet the minimum monthly basic electricity requirements per household of its non-electrified population is then estimated. The comparison between this energy needed to meet the minimum BE requirements and the actual energy generated is shown in Table 9 below. It is evident from the table that, for the countries with low electricity access rates, more energy is required to meet the minimum BE of 50kWh a month than the current energy generated by those countries. Ideally, this should be done through renewable energy to reduce the level of pollution associated with fossil-based generation, as discussed in Section 3.

Table 9: Estimated Annual Generation (GWh) needed to meet BE Requirements

Country	Total Population (UNData, 2013)	Fraction Without Electricity Access	Existing Energy Generated (GWh)	BE=50 kWh/mo (GWh/yr)	BE=100 kWh/mo (GWh/yr)	BE=200 kWh/mo (GWh/yr)
Angola	21,471,618	0.7	5,613	9,018	18,036	36,072
Botswana	2,021,144	0.34	250	412	825	1,649
D. R. Congo	67,513,677	0.91	7,966	36,862	73,725	147,450
Lesotho	2,074,465	0.72	486	896	1,792	3,585
Madagascar	22,924,851	0.85	2,025	11,692	23,383	46,767
Malawi	16,362,567	0.91	2,180	8,934	17,868	35,736
Mauritius	1,258,653	0	2,792	0	0	0
Mozambique	25,833,752	0.61	15,166	9,455	18,910	37,821
Namibia	2,303,315	0.7	1,643	967	1,935	3,870
Seychelles	89,173	0.03	320	2	3	6
South Africa	53,157,490	0.15	257,919	4,784	9,568	19,137
Swaziland	1,249,514	0.73	425	547	1,095	2,189
Tanzania	49,253,126	0.76	5,795	22,459	44,919	89,838
Zambia	14,538,640	0.74	11,850	6,455	12,910	25,821
Zimbabwe	14,149,648	0.6	9,088	5,094	10,188	20,375
Total	**294,201,633**		**323,518**	**117,579**	**235,157**	**470,315**

COST ANALYSIS: HOMER MODELLING

The general purpose of this section is to identify the most cost-effective technology options for off-grid distributed electrification, namely wind and solar power, in southern Africa when compared to both diesel generator and grid-based electricity. A previous study by Szabó, *et al.* (2011) used different optimisation tools than the one used here to map and estimate the least-cost off-grid electrification from solar power or diesel in all of Africa when compared to grid-based electricity. However, the present study not only includes wind power, it focuses on the meteorological conditions in southern Africa, and uses the most recent generation and cost data available for the region. It also reveals which alternatives to the grid are environmentally preferable with respect to their level of emission of pollutants even though no prices are placed on these in this study.

The main tool that was used for the analysis of the costs of the distributed renewable energy options described in this chapter is the Hybrid Optimization Model for Electric Renewables (HOMER), which was developed by the US National Renewable Energy Laboratory (NREL). HOMER is a micro-power optimisation model that simplifies the evaluation of both off-grid and grid-connected power system designs for various applications (NREL, 2005). It works by performing three main tasks: simulation, optimisation and sensitivity analyses.

The software first simulates the operation of a given system configuration by making energy balance calculations for each of the 8,760 hours in the year, and then determines whether the configuration is possible. It does this by verifying whether it can meet the electricity demand under specified conditions and by estimating the installation and operating costs over the lifetime of the project, which include capital costs, replacement costs, operation and maintenance costs, fuel and interest. HOMER optimises the system by simulating all its possible configurations and displays them as a list of the lowest to the highest net present cost, which it also defined as the life cycle cost, in order to compare different design alternatives. The software also performs a sensitivity analysis by repeating the optimisation process for each sensitivity variable that is specified, such as a range of wind speeds or a range of diesel prices.

The cost analysis in this section evaluates four scenarios as follows: 1) Diesel only; 2) Solar power versus diesel; 3) Wind power versus diesel; 4) Hybrid solar-wind power versus diesel. The analyses that are presented in this section include a description of all the technology and cost assumptions that are made, a display of the net present cost (life cycle cost) of the optimised technology options, the levelised cost of electricity for each energy resource, and a display of the emissions released by each of the resource options. The rand costs of components and fuels in South Africa are relied on since these are available online (Sustainable.co.za, 2015) and because that is the currency that the SAPP uses in its analyses. Thus, it is assumed that the results presented can be generalised for the SADC region in terms of purchasing power parity (PPP).

Table 10: Technical Specifications and Costs of Selected Components of Energy Systems

	Capacity	Capital Costs	Fuel Costs	Lifetime
Solar PV	250W	R3,009	R0	20 years
Wind turbine	400kW	R16,006	R0	15 years
Diesel Generator	4kW	R25,400	R11.318/L	15,000 hrs
Deep Cycle Battery	12V, 26 Amp-hr	R743	N/A	5 years
Inverter	12V, 150W	R389	N/A	

Simulation

A sample schematic of the solar-wind-diesel hybrid scenario that was conducted in HOMER is shown in Figure 2 below where Generator 1 represents the diesel generator, and the primary load represents a basic electricity load of 200kWh/month (or 6.7kWh/d). The icon with 'S6CS25P' label represents a deep-cycle battery, while the icons under 'Resources' allow the modeller to import characteristics such as average daily wind speed for every month of the year, the average daily solar radiation per month, modify the average daily amount of biomass used per month (tonnes/day) and the average price of the biomass used ($/tonne), and enter the average fuel price ($/L) in the case of diesel.

For the icons under 'Other', the user can modify parameters such as the annual real interest rate (for which six per cent is used here), the project lifetime, and/or the system capital and O&M costs. A converter icon is also shown because the diesel generator runs on electricity in an Alternating Current (AC) form, while the wind and solar generators run on a Direct Current (DC), which requires a converter in the form of an inverter (to convert DC to AC) or what is generically called a 'converter' to convert AC to DC. However, if household devices use a type of current that is the same as that of the generator being used, a converter is not needed.

Figure 2: Solar-Wind Hybrid versus Diesel

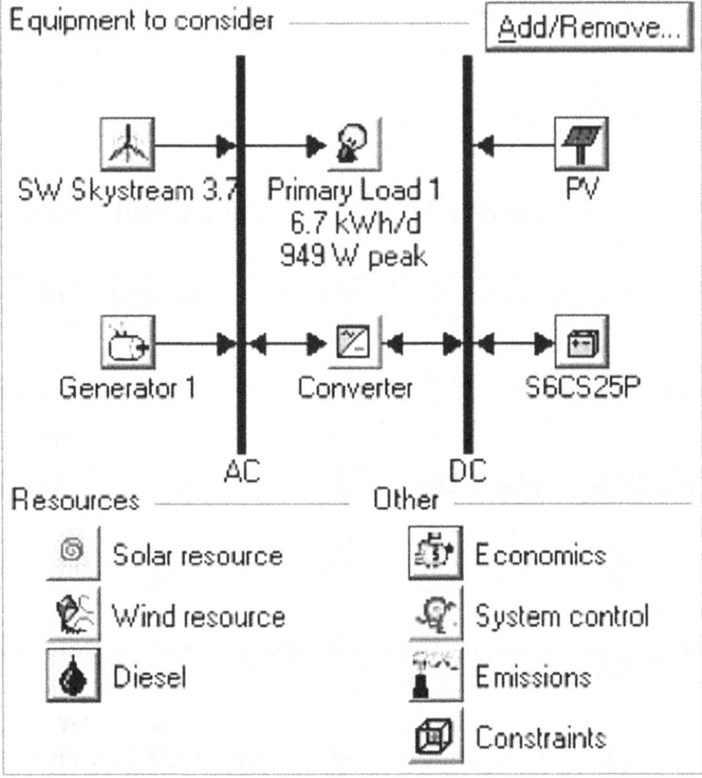

Optimisation and Sensitivity Analysis

The optimisation procedure that HOMER performs, which consists of simulating all the possible configurations of a system and ranking the most optimal configuration from the least expensive one to the most expensive, does not provide the 'solution' to the problem per se, but only allows the user to compare various alternative designs. However, it should be noted that optimisation itself, understood as the result of a formal objective function, helps to clarify given objectives under specific conditions such that learning can then occur over multiple iterations, as is typical of ecosystem management, for instance (Walters, 1986). Therefore, the modelling results presented in this section show the costs associated with the various alternatives prior to the social, technological and policy learning that will take place.

The total net present cost in the HOMER model is defined by Lambert, *et al.* (2006) as given by:

$$C_{NPC} = \frac{C_{ann,tot}}{CRF\ (i, R_{proj})}$$

where $C_{ann,tot}$ is the total annualised cost, i is the discount rate, R_{proj} is the project lifetime, and CRF(.) is the capital recovery factor, which is given by:

$$CRF(i,\ N) = \frac{i(1+i)^{N}}{(1+i)^{N}-1}$$

where i is the annual real interest rate and N is the number of years.

The levelised cost of energy (LCOE) is then given by:

$$COE = \frac{C_{ann,tot}}{E_{prim} + E_{def} + E_{grid,sales}}$$

where $C_{ann,tot}$ is the total annualised cost, E_{prim} are the total amounts of primary load, E_{def} represents the deferrable load that the system can generate per year, while $E_{grid,sales}$ is the amount of energy sold to the grid per year (Lambert, *et al.* 2006). In this case, $E_{grid,sales}$ is zero since no sales either to or from the grid were included in the model.

The results of the optimisation for each set of loads, for example, 50kWh/month basic electricity requirement (1.67kWh/day), while the sensitivity analysis runs the optimisation procedure again for 100kWh/month, 200kWh/month, 400kWh/month and 800kWh/month as shown in Tables 12 to 15 below for all four scenarios. The sensitivity analysis feature in HOMER most closely models the concept of component sizing, which has been used in this chapter to capture the possibility of gradually and adaptively building up electricity generation infrastructure. Thus, the main purpose of the sensitivity analysis was to determine what combination of the components could produce the daily equivalent of the monthly primary loads of 50kWh, 100kWh, 200kWh, 400kWh and 800kWh, and to subsequently observe the evolution of the levelised cost of electricity (LCOE) with an increase in consumption.

In Table 11 overleaf, for the diesel-only scenario, the labels in order are as follows: The Diesel (kW) tab represents the installed capacity of the diesel generator, followed by the initial capital required to install it, the operating costs per year, the total Net Present Cost (NPC) or life cycle cost, the levelised cost of electricity (LCOE), and the fraction of renewable energy in the resource used.

Tables 12, 13 and 14 also show the components of the PV and wind turbine systems, and display the net present costs of all the feasible configurations.

Table 11: Diesel only

Pri. Load (kWh/d)	Diesel (kW)	Initial capital	Operating cost (R/yr)	Total NPC	LCOE (R/kWh)	Renewable fraction
1.667	0.65	R4,128	12,154	R159,492	20.51	0
3.333	0.65	R4,128	12,154	R159,492	10.26	0
6.667	0.65	R4,128	14,181	R185,409	5.96	0
13.333	0.65	R4,128	20,981	R272,337	4.38	0
20	1.3	R8,255	35,160	R457,723	4.91	0
26.667	1.3	R8,255	41,966	R544,721	4.38	0

Table 12: Solar power versus diesel

Pri. Load (kWh/mo)	PV (kW)	Diesel (kW)	Initial capital	Total NPC	LCOE (R/kWh)	Renewable fraction
50	0.5	0.65	R14,993	R20,707	2.66	0.93
100	1	0.65	R21,011	R31,421	2.02	0.92
200	2	0.65	R33,047	R53,952	1.73	0.92
400	5	0.65	R73,613	R100,134	1.61	0.97
600	5	1.3	R77,740	R157,816	1.69	0.87
800	10	1.3	R146,836	R199,749	1.61	0.97

Table 13: Wind power versus diesel

Pri. Load (kWh/d)	Wind turbine	Diesel (kW)	Initial capital	Total NPC	LCOE (R/kWh)	Renewable fraction
50	1	0.65	R24,981	R70,352	9.05	0.14
100	1	0.65	R24,981	R113,904	7.32	0.07
200	1	0.65	R24,981	R187,547	6.03	0.04
400	1	0.65	R24,981	R297,474	4.78	0.02
600	1	1.3	R29,108	R472,259	5.06	0.01
800	1	1.3	R29,108	R569,826	4.58	0.01

Table 14: Solar-Wind hybrid power versus diesel

Pri. Load (kWh/d)	PV (kW)	Diesel (kW)	Initial capital	Total NPC	LCOE (R/kWh)	Renewable fraction
50	0.5	0	R26,871	R35,628	4.58	1
100	1	0.65	R37,017	R50,314	3.24	0.95
200	2	0.65	R49,053	R71,628	2.30	0.94
400	5	0.65	R89,619	R120,528	1.94	0.97
600	5	1.3	R93,746	R177,852	1.91	0.87
800	10	1.3	R162,842	R220,011	1.77	0.97

The tables above show that the levelised cost of electricity (LCOE) is reduced as consumption increases. This demonstrates that the economies of scale are clearly beneficial over the life cycle of the respective energy generation technologies. However, only the solar-diesel scenario shows a significant uptake of the renewable energy component (minimum 87 per cent, maximum 97 per cent). The wind-diesel scenario still relies on diesel to keep the cost at a minimum, while the solar-wind-diesel scenario also primarily relies on solar power to keep the cost at a minimum. Thus, over the average area of the various countries modelled, solar power is the most cost-effective resource across the region. What this suggests is that, outside of specifically well-resourced areas with respect to wind energy, household residents are not likely to benefit from this resource as much as solar, which is virtually ubiquitous in the region. Table 15 and Figure 3 below show the evolution of the cost of electricity for the four hybrid scenarios with increased consumption.

Table 15: Levelised Cost of Electricity (LCOE) across Four Hybrid Scenarios

Primary Load (kWh)	Diesel (R/kWh)	Solar-diesel (R/kWh)	Wind-diesel (R/kWh)	Solar-wind-diesel (R/kWh)
50	20.5 (0.00)	2.66 (0.93)	9.05 (0.14)	4.58 (1.00)
100	10.27 (0.00)	2.02 (0.92)	7.32 (0.07)	3.24 (0.95)
200	5.96 (0.00)	1.73 (0.92)	6.03 (0.04)	2.30 (0.94)
400	4.39 (0.00)	1.61 (0.97)	4.78 (0.02)	1.94 (0.97)
600	4.91 (0.00)	1.69 (0.87)	5.06 (0.01)	1.91 (0.87)
800	4.38 (0.00)	1.61 (0.97)	4.58 (0.01)	1.77 (0.97)

Note: The numbers in brackets reflect the fraction of renewable energy in each scenario.

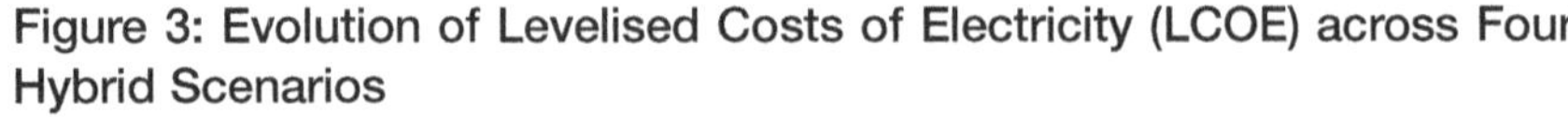

Figure 3: Evolution of Levelised Costs of Electricity (LCOE) across Four Hybrid Scenarios

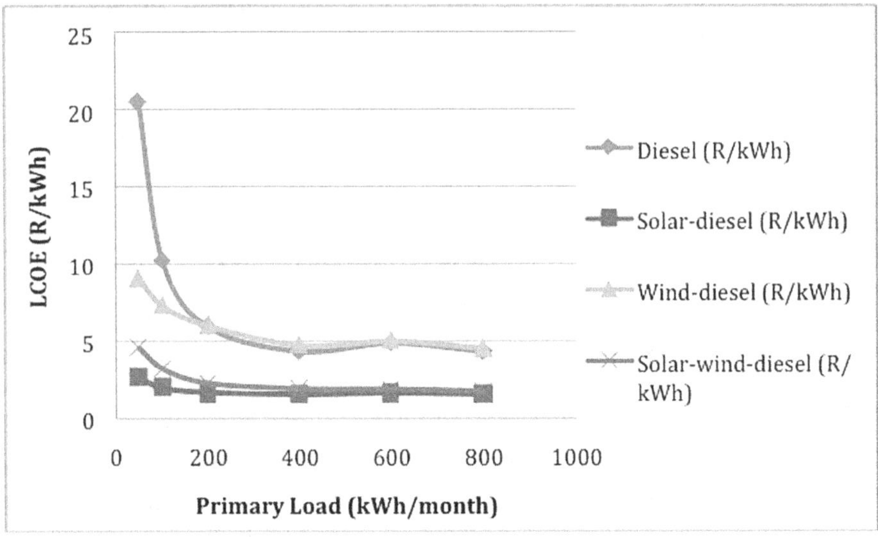

Two important points can be gleaned from the analyses presented above. In the first instance, the costs of the various scenarios are still higher than the average costs for grid-connected consumers in SADC, with the average consumer in Swaziland paying 1.37 R/kWh, those in the DRC paying 0.57 R/kWh and those in South Africa paying 1.03 R/kWh (SAPP, 2014). It should be noted, however, that this average for residential, commercial and industrial consumers conceals the fact that the industrial consumers often pay 'bulk rates' that are in the order of 50 per cent of the cost that other consumers pay, as in South Africa, for instance, although lower level consumers (less than 200 kWh/month) in that country are also given preferential rates (Eskom, 2014). Secondly, the increased consumption of the distributed, renewable technologies (solar, in this case) beyond what was modelled could very well benefit from economies of scale if larger neighbourhoods or communities were considered. Nevertheless, Table 16 below proceeds to estimate the capital costs that non-electrified households in southern Africa would need to invest to gain access to 50kWh per month of solar power. The assumption that was made was for five persons per household, and a capital cost of R10,865 (the difference between the initial capital cost of the solar-diesel scenario in Table 12 and that of the diesel-only scenario in Table 11, which leaves us with the capital cost of solar-only) was used.

Table 16: Estimated Costs of Extending Electrification through Renewable Energy

Country	Total Population (UNData, 2013)	Fraction without electricity access	Households without electricity access	Capital cost (BE=50kWh)
Angola	21,471,618	0.7	3,006,027	32,660,478,140
Botswana	2,021,144	0.34	137,438	1,493,261,610
D. R. Congo	67,513,677	0.91	12,287,489	133,503,570,310
Lesotho	2,074,465	0.72	298,723	3,245,624,960
Madagascar	22,924,851	0.85	3,897,225	42,343,346,040
Malawi	16,362,567	0.91	2,977,987	32,355,830,863
Mauritius	1,258,653	0	0	0
Mozambique	25,833,752	0.61	3,151,718	34,243,413,289
Namibia	2,303,315	0.7	322,464	3,503,572,447
Seychelles	89,173	0.03	535	5,813,188
South Africa	53,157,490	0.15	1,594,725	17,326,683,866
Swaziland	1,249,514	0.73	182,429	1,982,091,563
Tanzania	49,253,126	0.76	7,486,475	81,340,552,526
Zambia	14,538,640	0.74	2,151,719	23,378,423,893
Zimbabwe	14,149,648	0.6	1,697,958	18,448,311,062
Total	**294,201,633**		**39,192,911**	**425,830,973,756**

Table 16 above estimates the total cost of providing at least 50kWh/month to each household in southern Africa to be about R426 billion. For a sense of proportion, it is necessary to compare this figure to the $80 billion dollars (R950 billion) estimated for the construction of the Grand Inga project (Green, *et al.* 2015) or the estimated $17.8 billion (R211 billion) required for the 4,800MW Medupi Coal Power Station in South Africa, with its attendant controversies around the environment, labour and delayed completion (Rafey and Sovacool, 2011). Thus, this analysis suggests that the cost of electrifying the whole of southern Africa based on household systems could be less than 50 per cent of the cost of completing the Grand Inga Dam project and about twice the cost of building the Medupi Power Station. Given that such large infrastructure projects have a poor record of providing services to energy-poor consumers, and even when they do, this usually occurs a decade or more after their construction, the argument made for an investment in off-grid electrification is both technically practical, and economically feasible.

EMISSIONS

HOMER automatically calculates the amount of the various emissions that are released from the use of various resources as part of the optimisation procedure. Table 18 below shows the emissions result for all four hybrid scenarios for a household that consumes 200kWh/month, and a similar result can also be retrieved for the other loads. It should be noted that these results represent emissions that are proportional to the fractions of renewable versus diesel in all the hybrid scenarios that were run as shown in the optimisation results in Tables 11 to 14. The diesel generator emits the largest amount of CO_2 as well as all the other pollutants, followed closely by the wind-diesel scenario because only very little wind uptake was recorded due to the cost of wind technology and the relatively low wind resource in the areas surveyed. The solar-diesel scenario and the solar-wind-diesel scenario (mainly due to the solar component) demonstrate the significant emissions reductions that can be derived from the adoption of renewable energy.

Table 17: Estimated Emissions of Hybrid Scenarios at BE of 200kWh/month

	Diesel	Solar-diesel	Wind-diesel	Solar-wind-diesel
% Renewable Resource	0%	92%	4%	94%
Carbon dioxide (kg/yr)	3,539	283	3,397	212
Carbon monoxide (kg/yr)	8.73	0.70	8.38	0.52
Unburned hydrocarbons (kg/yr)	0.97	0.08	0.93	0.06
Particulate matter (kg/yr)	0.66	0.05	0.63	0.04
Sulphur dioxide (kg/yr)	7.11	0.57	6.83	0.43
Nitrogen oxides (kg/yr)	77.9	6.23	74.78	4.67

SUMMARY OF ANALYSES

In conclusion, this chapter has sought to demonstrate the technological and economic feasibility of power planning in southern Africa through off-grid electrification using suitable renewable energy resources, as a viable alternative to the centralised generation approaches that are dominant in the region. The study began with a presentation of the energy access situation in the region, the currently used resources for electricity generation, and the energy demand and supply in the SADC region. The first part of the analysis was based on a literature review of environmental assessments of various energy generation technologies on a life cycle basis in order to demonstrate the overwhelming evidence in favour of renewable energy technologies, not only in terms of harmful emissions at the point of use, but over the lifetime of the various devices from manufacture to disposal. The second part of the analysis estimated the electricity generation potential from both the solar and wind energy resources available in the various countries, and showed that they are sufficient to meet the energy demand of each of the countries many times over. Furthermore, the analysis showed how the various technological components needed for solar and wind power could meet the electricity of households needing to consume 50kWh/month, 100kWh/month or 200kWh/month.

Once the practical potential of both the renewable resources and the possibility of currently available technologies to exploit them were demonstrated, an analysis of the cost of these sources of electricity for distributed electrification was undertaken. The analysis showed that on a life cycle basis, the levelised costs of electricity (LCOE) for all the scenarios evaluated were higher than the average cost of grid-based electrification. In the case of the diesel-only scenario, which is the traditional alternative to grid-based electricity, it was shown that the recurring cost of the diesel fuel made this option many times more expensive than grid-based electricity even at higher consumption levels.

On the other hand, the solar-diesel scenario, which demonstrated a high uptake (> 90%) of the solar resource, showed that its LCOE was consistently declining and was not much more expensive than grid-based electricity at consumption levels greater than 200kWh/month. However, the cost of the wind-diesel scenario was expensive due to the relatively high cost of the wind generation technology and the modest energy resource over the areas surveyed. The solar-wind-diesel scenario was also favourable, due to the cost-effective solar resource and a slight contribution from wind. Thus, this part of the analysis showed that of the two renewable resources, solar power is the preferred option from both a resource and a cost perspective.

It should be noted that the bulk of the costs of the renewable energy scenarios was in the capital costs, since the fuel costs are non-existent. Thus, the costs of supplying the initial equipment for generating 50kWh/month per household in southern Africa were estimated, which is consistent with the Free Basic Electricity (FBE) policy in South Africa that was responsible (in conjunction with institutional and technical innovations) for its rapid electrification programme. The results suggest that the total costs needed to provide every non-electrified household in the region with a minimum of 50 kWh/month of solar power would represent less than 50 per cent of the estimated cost of completing the Grand Inga Dam and about twice the cost of one power station, the Medupi Coal Power Station. Given that the residential consumers in South Africa represented approximately 94 per cent of Eskom's customers and used about 20 per cent of the total electricity generated in 2006 (National Treasury, 2011), then the following proposition can be considered. In the event of an ambitious regional electrification programme, the prospect that only two times the allocated costs for new generation at one South African power plant could supply power to all the non-electrified residents in the wider region (or half of them, if a quarter of the generating capacity of each of the two plants were dedicated to residential consumers) and within a much shorter period of time, remains compelling. This strongly suggests that the generalised neglect or deficiency of state-based initiatives for widespread electrification in sub-Saharan Africa in general, and in southern Africa more specifically (with few exceptions such as South Africa with respect to grid expansion, and some of the island nations), cannot continue to be justified on the basis of cost while simultaneously championing the development of mega projects and restructuring efforts that, since the early 1980s, have only made marginal contributions to either electricity access or reliability of supply (Turkson, 2000) – a situation that appears to continue to deteriorate. More fundamentally, the overwhelming support for rural electrification in countries that have achieved near-universal access such as China, Tunisia or Mexico (Barnes, 2007), and empirical support for the link between off-grid electrification and human development in Brazil's Luz para Todos (Light for All) programme (Gómez and Silveira, 2010), for instance, all point to the soundness of such an approach.

In addition, it should be pointed out that the scenarios investigated in this study by no means exhaust the opportunities for off-grid electrification, with biomass power from waste crops being one option that would be especially suited to farming communities in rural areas. For instance, a modelling-based assessment of hybrid PV-fuel cell systems coupled with batteries by Lagorse, *et al.* (2009) demonstrates its potential for an off-grid application related to stand-alone street lighting systems, but the study's findings underscore the fact that such an option is particularly attractive for locations that are distant from the equator and where PV-battery systems cannot work all year round. Thus, while South Africa, in particular, with its rich platinum reserves, has the potential to develop proton exchange membrane fuel cells for off-grid applications (Ferreira and Perrot, 2013), the reality of the region's generally exceptional solar radiation suggests that this technology is not likely to be a cost-competitive solution in the short-to medium-term. Lastly, the scenario analysis in this chapter has shown that the renewable energy scenarios (particularly the two that included solar power) would result in less pollution at the point of use than in the diesel-based scenario. Thus, if one considers the extended delays that are typically associated with large power generation infrastructure as well as the environmental ramifications of both fossil-fuelled generators and large hydroelectric dams, along with the cost and debt implications of constructing those, the combined analyses presented in this chapter show convincingly that power planning based on off-grid electrification deserves serious contemplation in the near future.

RECOMMENDATIONS

While the analyses offered in this chapter may be revised and some of its assumptions may be subjected to critique, the objective was to make a credible case for an alternative planning paradigm based on the magnitude of the demand, resources and costs of the available technologies. A few additional recommendations for implementing such a paradigm are offered as follows:

Financing

The current study has pointed out that sub-Saharan African countries and regions have been able to generate the capital needed for large energy infrastructure projects with highly skewed distributional consequences and that the approach suggested here would only represent a very small percentage of those costs even if the capital costs of the off-grid technologies were entirely paid for. Other approaches to financing off-grid electrification would be other forms of subsidies (partial), remittances, rotating credit and savings associations, village and cooperative savings and credit institutions, or investment funds for manufacturing and research, as discussed in more detail in a study by Soumonni and Soumonni (2011). With specific reference to southern Africa, the Integrated National Electrification Plan (INEP) of the Department of Energy (DoE) of South Africa supports a rural concessions programme for solar home systems by providing 80 per cent of the capital costs, with the one run by the Nuon-RAPS (NuRA) private utility since 1999 having achieved notable success in KwaZulu-Natal (Lemaire, 2007).

Technological Capability Building

The technical capability needed to install and maintain the energy generation technology is minimal at the moment, but this represents an opportunity for training young people at all post-secondary levels including technical and artisan training institutes, as well as traditional polytechnics and universities. Adopting this paradigm of electrification would provide an opportunity to foster an approach to education that is more directly linked to pressing developmental issues, while simultaneously creating opportunities for entrepreneurship and work opportunities.

Industrial Consumers

It has been suggested in this study that large infrastructure projects tend to favour large industrial consumers that are predominantly linked to the extractive sectors of the economy. Yet, such consumers have the financial means and the opportunity to also explore opportunities for distributed electrification through co-generation or Combined Heat and Power (CHP). If the preferential electricity supply arrangements for such consumers were more equitably revised with respect to other societal groups, they may find that the possibilities of generating a large proportion of their own power may be cost-effective and competitive with the current prices they pay. In addition, for the various countries, promoting such an approach through policy instruments could both stimulate manufacturing (i.e. non-extractive industry), promote associated job growth and negate the need for building expensive and ecologically compromised large power stations, as suggested in a number of recent studies (Brown, *et al.* 2013, Baer, *et al.* 2015).

Grid-connected Residential and Commercial Consumers

In many countries in southern Africa, many grid-connected consumers are experiencing power cuts or load shedding, including in countries that have historically had access to an abundant energy supply, such as South Africa. This recent development has tended to relegate issues of electricity access to the background, in favour of the reliability of supply for grid-connected consumers. Given the indications that this situation may continue for at least a few years, it is possible for such consumers to adopt distributed renewable electricity to supplement their current supply, thereby mitigating the need for additional centralised generation capacity, as well as ensuring the reliability of their own power supply. These consumers could ultimately be connected with those who started out as off-grid users in order to develop in the future what has been proposed in the US as a distributed utility (Feinstein, *et al.* 1997), but which may be much more applicable to the socio-technical conditions in Africa.

Innovation

Ultimately, the goal of moving from resource-dependent economies to knowledge-driven economies must revolve around solving the problems that affect the majority of African people. In the case of electricity access, adopting an off-grid paradigm based on renewable energy could provide opportunities to integrate the technological learning and capability that would be gained through adoption, with the ongoing research and development efforts at university and governmental research centres. This would enable the countries in southern Africa as well as other sub-Saharan African countries to develop a range of innovation capabilities in the area of renewable energies that are tailored to their own environments, and that can subsequently serve as a basis for a future export industry to other parts of the world. Such considerations for the continent from an innovation systems perspective have been further developed in a study by (Soumonni, 2013).

Appendix 1: Solar Energy: Geographical Coordinates and Average Radiation Data (NASA, 2015)

Country	City	Latitude	Longitude	Elevation (m)	Average resolved Solar Radiation (kWh/m2day)
Angola	Luanda	13.26	8.83	59	--
	Cuando Cubango	18.8	16.33	1,221	6.54
Botswana	Gaborone	25.89	24.65	1,159.26	--
	Maun	23.42	19.98	942	6.36
D. R. Congo	Lubumbashi	27.5	11.67	1,303	--
	Kinshasa	15.32	4.33	351	6.19
Lesotho	Maseru	29.31	31.98	1,642	--
	Mokhothong	29.07	29.29	2,209	5.81
Madagascar	Antananarivo	47.53	18.93	1,352	--
	Itampolo	43.95	24.68	51	3.49
Malawi	Lilongwe	33.77	13.98	1,135	--
	Kaphiika	34.09	10.45	692	5.33
Mauritius	Port Louis	57.52	20.17	85	--
	Mahebourg	57.7	20.4	44	3.17
Mozambique	Maputo	32.56	26.01	22	--
	Tete	33.6	16.17	270	5.54
Namibia	Windhoek	17.11	22.6	1,723	--
	Okahandja	16.92	21.94	1,578	6.38
Seychelles	Victoria	55.45	4.62	72	--
	La Digue	55.83	4.36	206	3.02
South Africa	Cape Town	18.42	33.9	-1	--
	Thohoyandou	30.48	22.95	576	6.19
Swaziland	Mbabane	31.13	26.32	1,335	--
	Hluti	31.59	27.22	519	5.29
Tanzania	Dar es Salaam	39.28	6.78	23	--
	Mwanza	32.9	2.51	1,187	4.95
Zambia	Lusaka	28.28	15.43	1,185	--
	Kasama	31.2	10.23	1,365	5.57
Zimbabwe	Bulawayo	28.62	20.12	1,211	--
	Harare	31.03	17.85	1,351	6.54

REFERENCES

Africa, E. 2010. 'Free Basic Electricity: A Better Life for All'. Johannesburg, South Africa: Sustainable Energy & Climate Change Project, Earthlife Africa.

Baer, P., Brown, M. A. and Kim, G. 2015. 'The job generation impacts of expanding industrial cogeneration', *Ecological Economics*, 110, 141–153.

Barnes, D. F. 2007. 'The Challenge of Rural Electrification'. In: Barnes, D. F. (ed.) *The Challenge of Rural Electrification: Strategies for Developing Countries*. Washington, DC: RFF Press.

Bekker, B., Eberhard, A., Gaunt, T. and Marquard, A. 2008. 'South Africa's rapid electrification programme: Policy, institutional, planning, financing and technical innovations', *Energy Policy*, 36, 3115–3127.

Borbely, A.-M. and Kreider, J. F. 2001. 'Distributed Generation: An Introduction'. In: Borbely, A.-M. & Kreider, J. F. (ed.) *Distributed Generation: The Power Paradigm of the New Millenium*. Boco Rotan: CRC Press.

Brown, M., Cox, M. and Baer, P. 2013. 'Reviving Manufacturing with a federal cogeneration policy', *Energy Policy*, 52, 264–276.

Byrne, J., Zhou, A., Shen, B. and Hughes, K. 2007. 'Evaluating the potential of small-scale renewable energy options to meet rural livelihood needs: A GIS and lifecycle cost-based assessment of Western China's options', *Energy Policy*, 4391–4401.

Cherubini, F., Bird, N. D., Cowie, A., Jungmeier, G., Schlamadinger, B. and Woess-Gallasch, S. 2009. 'Energy- and greenhouse gas-based LCA of biofuel and bioenergy systems: Key issues, ranges and recommendations', *Resources, Conservation and Recycling*, 434–447.

Coley, D. 2008. *Energy and Climate Change: Creating a Sustainable Future*. West Sussex: Wiley.

DME. 2007. 'Free Basic Alternative Energy Policy (Household Support Programme)'. South Africa: Department of Minerals and Energy.

Eskom 2014. 'International Benchmarking of Electricity Tariffs'. Research Report.

Feinstein, C. D., Orans, R. and Chapel, S. W. 1997. 'The Distributed Utility: A New Electric Utility Planning and Pricing Paradigm', *Annual Review of Energy and the Environment*, 22, 155–185.

Ferreira, V. and Perrot, R. 2013. 'Innovation and Hydrogen Fuel Cell Manufacturing'. In: MISTRA (ed.) *South Africa and the Global Hydrogen Economy: The Strategic Role of Platinum Group Metals*. Johannesburg: Real African Publishers.

Fleck, B. and Huot, M. 2009. 'Comparative life-cycle assessment of a small wind turbine for residential off-grid use', *Renewable Energy*, 2688–2696.

Gauthier, F., Lubes-Niel, H., Sabatier, R., Masson, J. M., Paturel, J. E. and Servat, E. 1998. 'Variabilité du regimé pluviométrique de l'Afrique de l'Ouest non sahélienne entre 1950 et 1989', *Hydrological Sciences-Journal-des Sciences Hydrologiques*, 43, 15.

Gómez, M. F. and Silveira, S. 2010. 'Rural electrification of the Brazilian Amazon – Achievements and lessons', *Energy Policy*, 38, 6251–6260.

Green, N., Sovacool, B. K. and Hancock, K. 2015. 'Grand Designs: Assessing the African Energy Security Implications of the Grand Inga Dam', *African Studies Review*, 48, 133–158.

Hafemeister, D. W. 2007. *Physics of Societal Issues: Calculations on National Security, Environment, and Energy*. New York: Springer.

IEA. 2012. IEA Statistics. Available at: http://www.iea.org/statistics/statisticssearch/ [Accessed 20 July 2015].

IEA. 2014. World Energy Outlook 2014 Electricity Database – Electricity Access in Africa in 2012. Available at: http://wwwworldenergyoutlook.org/resources/energydevelopment/energyaccessdatabase/ [Accessed 20 July 2015].

Lagorse, J., Paire, D. and Miraoui, A. 2009. 'Sizing optimization of a stand-alone street lighting system powered by a hybrid system using fuel cell, PV and battery', *Renewable Energy*, 34, 683–691.

Lambert, T., Gilman, P. and Lilienthal, P. 2006. 'Micropower System Modeling with Homer'. In: Farret, F. A. and Simões, M. G. (eds.) *Integration of Alternative Sources of Energy*. Hoboken, New Jersey and Canada: John Wiley & Sons, Inc.

Lemaire, X. 2007. 'Concession for rural electrification with solar home systems in KwaZulu-Natal (South Africa)'. Centre for Management under Regulation – Sustainable Energy Regulation Network, Warwick Business School.

Levin, T. and Thomas, V. M. 2011. 'Least-cost network evaluation of centralized and decentralized contributions to global electrification', *Energy Policy*, 41, 286–302.

NASA. 2015. NASA Surface meteorology and Solar Energy: Daily Averaged Data. Available at: http://eosweb.larc.nasa.gov/ [Accessed 20 July 2015].

National Treasury. 2011. 'Local Government Budgets and Expenditure Review: 2006/07–2012/13'. South Africa.

Nieuwlaar, E. 2004. 'Life Cycle Assessment and Energy Systems', *Encyclopedia of Energy*, 3, 647–654.

NREL. 2005. Getting Started Guide for HOMER Version 2.1. Available at: http://homerenergy.com/user_interface.html [Accessed 20 July 2015].

Pascale, A., Urmee, T. and Moore, A. 2011. 'Life cycle assessment of a community hydroelectric power system in rural Thailand', *Renewable Energy*, 2799–2808.

Pineau, P.-O. 2008. 'Electricity sector integration in West Africa', *Energy Policy*, 36, 210–223.

Rafey, W. and Sovacool, B. K. 2011. 'Competing discourses of energy development: The implications of the Medupi coal-fired power plant in South Africa', *Global Environmental Change*, 21, 1141–1151.

SAPP. 2014. Annual Report. Available at: http://www.sapp.co.zw/docs/Annual%20report-2014.pdf [Accessed 20 July 2015].

SAPP. 2015. Vision and Objectives. Available at: http://www.sapp.co.zw/ [Accessed 20 July 2015].

Sobin, R. 2007. 'Energy Myth Seven – Renewable Energy Systems Could Never Meet Growing Electricity Demand in America'. In: Sovacool, B. K. and Brown, M. A. (eds.) *Energy and American Society – Thirteen Myths*. Dordrecht: Springer.

Soumonni, O. 2013. 'Towards a Technology Policy for Renewable Energy Development in Africa: A Systems of Innovation Perspective', *African Journal of Science, Technology, Innovation and Development*, 5, 289–295.

Soumonni, O. C. and Soumonni, O. Y. 2011. 'Promoting West African Ownership of the Power Sector: Alternative Financing for Distributed Generation of Renewable Electricity', *Journal of African Business*, 12, 310–329.

Sovacool, B. 2008. 'Distributed Generation in the US – Three Lessons', *Cogeneration and On-Site Power Production*, 69–72.

SUSTAINABLE.CO.ZA. 2015. SA's No. 1 Online Eco Store. Available at: http://www.sustainable.co.za/ [Accessed 20 July 2015].

Szabó, S., Bódis, K., Huld, T. and Moner-Girona, M. 2011. 'Energy solutions in rural Africa: mapping electrification costs of distributed solar and diesel generation versus grid extension', *Environmental Research Letters*, 6, 1–9.

Turkson, J. K. 2000. *Power Sector Reform in sub-Saharan Africa.* Houndmills, Basingstoke, Hampshire and London: St. Martin's Press.

Walters, C. J. 1986. *Adaptive Management of Renewable Resources.* New York: Macmillan.

Weisser, D. 2007. 'A guide to life-cycle greenhouse gas (GHG) emissions from electric supply technologies', *Energy*, 1543–1559.

Wood, D. 2011. *Small Wind Turbines: Analysis, Design, and Application.* London, Dordrecht, Heidelberg and New York: Springer.

Zhong, Z. W., Song, B. and Loh, P. E. 2011. 'LCAs of a polycrystalline photovoltaic module and a wind turbine', *Renewable Energy*, 2227–2237.

SECTION IV

Summary of Policy Recommendations

Summary of Policy Recommendations

The various chapters of this book offer a number of policy recommendations and considerations towards a smooth transition to a low-carbon economy in South Africa, covering critical sectors of the green economy such as energy, transportation, waste and water.

SECTION I: THE CO-EVOLUTIONARY ROLE OF GOVERNMENTS, CIVIL SOCIETIES AND INDUSTRIES

Chapter 1: The Trojan Horses of Global Environmental and Social Politics

This chapter offers a critique of the dominant discourse on sustainable development, the green economy and climate change policies and examines how the hegemonic discourse around market-led economic growth principles, such as green growth, have led to decades of mitigation deadlocks and has further widened the gulf between the countries of the North and the South. A few considerations are critical in reframing the discussions around sustainable development and the green economy:

- There is a need to approach the issue of sustainable development in a manner that embraces the social complexity and the human and multidimensional nature of the issues. This requires a revolutionary change in political consciousness within and across national boundaries, including a keen appreciation of the conditions of the poor and marginalised across the globe; and the climate change and sustainable development policies of at least the major countries in the North and the South must be embedded within a broader transparent international effort and discourse.
- The world is in need of a climate change framework for transnational and global political action that embraces new and alternative thoughts

and action. Such thinking requires the spread of certain core values of sustainable development and the sharing of these values among countries.

- Alternative frameworks and modes of thinking and action will unlock technological, institutional and economic systems from old habits and path dependencies. Now is the time to scrutinise the specific ways in which the current world order has been created, the kind of thinking and practices that have led to its failings, and the systemic interventions required to extricate humanity from the current malaise.
- Both at the UNFCCC and at the national levels of the BASIC (Brazil, South Africa, India and China) countries, the important issue of adaptation as a climate change policy has been an under-discussed challenge. It has, however, gained ground since COP20 in December 2014, and features on the agenda of COP21 in December 2015. This must be sustained and intensified, taking into account the fact that the poor, especially in Africa, Asia and Latin America, are in the majority and are extensively dependent on climate-sensitive sectors for their sustenance.

Chapter 2: LTMS and Environmental and Energy Policy Planning in South Africa: Betwixt Utopia and Dystopia

The chapter argues that the LTMS and the energy and environmental policies it inspired have not been wholly effective. There are several reasons for the marked ineffectiveness of the LTMS and therefore a number of recommendations in this regard:

- The experience from other countries shows that policies rarely address climate change in isolation; rather, they are designed to fulfil a range of parallel objectives, be they energy security, reduced air pollution, economic restructuring or targeted industrial development.
- The big bottleneck is the implementation of policy interventions and sectoral strategies which is burdening not just the specific sector but the country at large through adverse multiplier effects. Policy synergies that integrate the social and economic priorities of the country, across and within government spheres, specifically between policies for climate change, industrial development, job-creation, poverty reduction and the energy sector are needed.

- The government needs to exploit 'critical junctures' such as the present electricity crisis to move beyond short- and medium-term vested interests as well as old policies and understandings and implement solutions that deliver results.
- In 'critical junctures', the structural constraints imposed on actors during the path-dependant phase are substantially relaxed. This provides an opportunity for policy entrepreneurs and other actors to reshape existing institutions and create new arrangements. Actors can influence outcomes and, where there are positive feedback effects, break the chain of path dependency and in so doing there is an opportunity to change the rules.
- The transition from the politically impossible to the politically inevitable compels us to re-imagine and rethink our policy 'utopia' and ensure that it is grounded in material remaking of economics and development. Reconnecting real world economics and transformative mitigation is the imperative right now.

Chapter 3: Historical Review on the Relationship Between Energy, Mining and the South African Economy

This chapter uses a co-evolutionary lens to discuss the (deep) challenges of a smooth transition to a low-carbon economy in South Africa. A number of recommendations and suggestions are made on how the complexity of the energy intensive challenge could be overcome and clean energy policies could thereby be effectively implemented:

- The role of the Minerals Energy Complex (MEC) is indicative of the entrenched regime that the mining sector plays within the economy. The MEC has broadened itself to a system of wealth accumulation and has integrated large sectors that include mining, transport, manufacturing and finance. Due to this system of wealth accumulation, there are a handful of players in this sector, and thereby also few powerful actors that are influential in determining key policies and strategies. An evaluation of the vested interests within the MEC is key in understanding the implementation of energy policies in the country.
- The issue of ideological mismatch between a developmental state and a mature market-based economy is a fundamental challenge. As such, issues such as social compacting in an environment in which one sector

is perceived to be strategically networked and more influential, the debates about the place and role of state-owned enterprises such as Eskom, and the ongoing discourse on nationalisation versus privatisation remain unresolved. The ideological mismatch has to be tackled because it is a root cause of policy incoherence and implementation inertia.

- The subject of financialisation is indicative of a need for major reform in the structure of the economy. Investments made by the mining sector that favour most profits flowing out of the country instead of long-term financial investments are compounding developmental challenges. These include uneven wealth distribution, mostly skewed towards elitism and towards wealth concentration, thus aggravating inequality. The JSE total market capitalisation is twice the size of the actual economy. This imbalance renders the economy vulnerable to short-term volatile capital flows, and not necessarily long-term labour absorbing economic growth. The current energy crisis provides an opportune moment to revisit the country's mining economic policy.

Chapter 4: Lost in Procurement: An Assessment of the Development Impact of the Renewable Energy Procurement Programme

The current design and the related practices of the REIPPP Programme in terms of maximising the socio-economic impact can be greatly improved. The analysis reveals 10 barriers that have hindered a positive impact on the development aspect of the sector. By expanding the interpretation of development, the programme can contribute the following to the national development agenda, through 'Seven Equations of Development':

- The creation of BEE or black-owned and run energy companies, which also implies the establishment of Black or South African industrialists.
- The transfer of skills from foreigners to locals in the most senior executive functions as well as the most junior roles entailed in constructing and operating RE power plants.
- The creation or growth of small enterprises that can leverage the experience of servicing power plants to participate in other sectors with similar needs.
- The development of communities into active agents in their own story of 'good change'.

- Impactful social investments owing to the participation of local communities in the articulation of development strategies and their resultant stewardship in managing the investments made in their communities.

SECTION II: TRANSITION TO A LOW-CARBON ECONOMY

Chapter 5: Making Transitions to Clean and Sustainable Energy in the South African Urban Transport Sector: Linkages to Growth and Inclusive Development

In the transition to a low-carbon economy, South Africa has been continually focusing on cleaner fuels as opposed to stimulating a move towards clean fuels. Some policy options to consider when moving from cleaner to clean technologies in the urban bus transport sector, and reducing the uncertainties that emerge under the conditions that characterise energy transitions and which might enable the movement towards change in the near future, are:

- Understanding habits, norms and practices that create lock-ins and path dependencies when making fuel choices, and learning to work with and around them is an important first step in an energy transition process.
- Making decisions about moving to new fuels would also require an impartial analysis of the costs that would be incurred and their likely evolution and the need for considerable collective action on the part of governments, users and producers working together.
- Considerable attention must be given to meeting the concerns of path dependence and lock-in, both of which are enhanced by weaknesses in carrying out policies and programmes that have been negotiated by stakeholders. One of the options is to bring into the choice process linkages among users and producers, as well as their perspectives on the nature of existing challenges and how these might be overcome. Closer linkages between users and producers is likely to enhance greater knowledge-sharing and transfer, including knowledge about the interchangeability between (new) technologies and a deeper understanding of the proximity of these technologies to full development and their use. Developing the capacity to work with and

around actors such as those from the petroleum, and oil and gas industry would be critical.

Chapter 6: Green Policymaking and Implementation at the City Level: Lessons from Efforts to Promote Commuter Cycling in Johannesburg

The development of a commuter cycling infrastructure and associated policy development in Johannesburg demonstrates the internal political dynamics in local government that allow such schemes to be championed and implemented, but which may also slow or stall progress. These processes do not happen independently of outside actors, which include a range of lobby groups that can serve to increase accountability or to divert policy towards parochial concerns. A number of considerations to improve the commuter cycling infrastructure and make the transition towards a low-carbon economy in the city are:

- The integration of bicycles into other forms of public transport, such as the trains and buses, is crucial to linking cyclists in far-flung suburbs with destinations in the city centre. While space constraints may prevent this in peak hours, a system could be devised to allow bicycles on buses, trains and the Gautrain during off-peak hours initially. This should be accompanied with safe bicycle parking at stations.
- With cycle lanes underway and nearing completion for many key routes, the other obstacle to greater uptake of cycling lies in bicycle access. Making bicycles and their repair affordable for low-income citizens should be a major priority of further work. Solutions could also be sought through partnership with the private sector.
- Given the large challenge posed to eco-mobility solutions such as cycling by the vast distances created by South Africa's incredibly low-density urban development, the commuter cycling cause must also be linked to demands to improve the public mass transit system and also for policies that encourage densification of existing settlements.
- Current cyclists, while not numerous, are a constituency who have been marginalised in policy and debate. More hard data should be collected about when, how often, and under what conditions people currently cycle. Data should also be collected and monitored about road accidents involving cyclists.

- Partnership between the state, the private sector, universities (and research institutes) and lobby groups should feed into the accountability and momentum for policy implementation, as well as in improving the content of policies by providing 'user' input.
- International networks, on the other hand, have had much more muted value. The value they have added in disseminating ideas and providing incentives for cities to excel as 'pioneers' in environmental innovation should be further examined. There is also recognition of the need for directing investment of time and personnel towards extending partnerships at the local level instead.

Chapter 7: The Energy and Water Nexus: The Case for an Integrated Approach for the Green Economy in South Africa

Some recommendations that would facilitate the cooperative and adaptive management of water and energy and thus help achieve the green economy goals for South Africa are:

- Better data availability and accuracy: improved availability and accuracy of data will facilitate informed decision-making, prioritise investments in both energy and water infrastructure and lead to better water and energy use practices.
- Integrated Resource Planning and Management: the relationship between water and energy needs to be better recognised in policies, planning and related regulations and laws dealing with the planning, management and development of water resources and energy systems.
- Efficiency Improvement: there is an urgent need to prioritise leakage reduction and energy efficiency in the water sector. Such measures include water pressure management, sludge management activities such as aeration efficiency improvement, using chemical pre-treatment (this could lead to potential electricity savings of 250–280 MWh/year), and optimisation of the operation of the distribution system by pumping at off-peak periods, among others.
- Appropriate resource pricing: appropriate pricing can play an important role in driving conservation and innovation in management and use of both resources through various pricing mechanisms.
- Increased funding for Research and Development (R&D) in smart technologies and innovations: there are many opportunities to develop

more efficient technologies, practise cost-effective approaches to using lower-quality, non-traditional sources of water to supplement or replace fresh water for cooling and other power plant needs. Future R&D should focus on: energy sources that can meet future water needs sustainably, specifically for the array of water-scarce areas in the country; technical solutions that successfully couple energy and water generation; reducing water use in thermal power generation through advanced cooling technologies, scrubbing, innovative source-water intake designs, the use of non-traditional waters and increased power-plant efficiencies, among others.

- Creating consumer awareness: urgent steps need to be taken by the government and the water companies to increase consumer awareness of water use. Tools such as the certification and labelling of consumer products to reflect embedded water and energy use in their manufacture or usage need to be mandated.

Chapter 8: Waste Re-use: Oil Extraction from Waste Tyres and Improvement of the Waste Tyre Industry

There is a need to assess the current waste tyre pyrolysis plants in South Africa in order to evaluate the economic and technical viability of these plants. Currently, the factors making these plants non-feasible, and around which relevant industrial policy interventions could be made, are:

- The complicated process of obtaining environmental permits to operate such plants.
- There is a need for beneficiation and sale of carbon black as a supplementary revenue stream.
- None of the South African facilities is utilising excess gas produced in plant to optimise energy utilisation.
- Localise technologies and ensure that they comply with relevant SABS codes.
- Improve the availability of spare parts, which are not readily available in the country.
- There is no guarantee of feedstock (waste tyres) and product off-take at pre-agreed prices and quantities.
- Most plants produce crude products of no value and require beneficiation, therefore consideration must be given to the beneficiation

of crude oil into valuable chemicals, which improves the plant process economics.

Chapter 9: Energy-efficient Low-income Housing Development in South Africa: The Next Build Programme

The policy framework to enable the national implementation of energy-efficient and renewable energy technology in the development of low-income houses exists in the country. However, national and provincial government departments have to coordinate policy action as some of the enabling legislation resides in different spheres and departments.

- Municipalities or city-level governments have to incorporate green energy development in their respective IDPs. The implementation of a national project on energy efficiency and renewable energy technology deployment for low-income housing has to be accompanied by continual monitoring and evaluation and this capacity has to be developed in all municipalities.
- Research organisations and universities will need to be involved in the green energy development programme to provide policy and decision-makers with informed evidence on viable energy solutions and services.
- The energy-efficient low-income housing and renewable energy development has the potential to transform the economy through innovation and development of new industries. Energy-efficient low-income housing development is an area where government has a captive market.

SECTION III: EXPLORING REGIONAL OPTIMAL ENERGY STRATEGIES

Chapter 10: Off-grid Renewable Electrification as a Viable and Complementary Power Planning Paradigm in Southern Africa: A Quantitative Assessment

A number of recommendations for implementing an alternative-planning paradigm based on the magnitude of the demand, resources and costs of the available technologies for off-grid solutions are as follows:

- Financing off-grid electrification could be in the form of subsidies (partial), remittances, rotating credit and savings associations, village and cooperative savings and credit institutions, or investment funds for manufacturing and research.
- Training young people at all post-secondary levels including technical and artisan training institutes, as well as traditional polytechnics and universities, would enhance the technical capabilities and skills base of the country. Education that is directly linked to pressing developmental issues would create opportunities for entrepreneurship and work opportunities for the youth.
- Industrial consumers with their financial means could explore opportunities for distributed electrification through co-generation or combined heat and power (CHP). There are possibilities of generating a large proportion of their own power, which may be cost-effective and competitive with the current prices. Promoting such an approach through policy instruments could stimulate manufacturing (i.e. non-extractive industry), promote associated job growth, and negate the need for building expensive and ecologically compromised large power stations.
- Strategies and policies should encourage residential and commercial consumers to adopt distributed renewable electricity to supplement their current supply, thereby mitigating the need for additional centralised generation capacity, as well as ensuring the reliability of their own power supply where existing access is erratic and short in supply.
- Adopting an off-grid paradigm based on renewable energy could provide opportunities to integrate the technological learning and capability that would be gained through adoption, with the ongoing research and development efforts at universities and governmental research centres. This would enable the countries in southern Africa as well as other sub-Saharan African countries to develop a range of innovation capabilities in the area of renewable energies that are tailored to their own environment, and which can subsequently serve as a basis for a future export industry to other parts of the world.

Index

H

I

J

K

L

M

T

U

V

W

Z

Printed in the United States
By Bookmasters